DANKMAR BOSSE

How Can Granite Arise from Basalt?

PART I

The Evolution *of* Granite

CHAPTER 1

The Four Minerals *of* Granite

To fully consider the main question of this book – how can granite arise from basalt – we must consider how the mineral world formed on the Earth (Part I), the formation of the Earth (Part II), and the separation of the nature kingdoms from the human being (Part III). In this chapter we will look at granite on which the living nature kingdoms and human beings were formed.

Granite consists of four minerals: mica, two feldspars, and quartz. Examples are shown in Figures 1 and 2. Before crystallizing, the minerals passed from a liquid through a state of pliability into a solid state. The black mica biotite separated first, allowing it to form its crystal shape most fully (Figures 3 and 4). This was followed by white feldspar (plagioclase), shown in Figures 5 and 6. Then came the red feldspar (orthoclase), shown in Figures 7 and 8. And finally, the gray, translucent quartz filled the remaining gaps in all of these pictures of granite as a last step (Figure 9).

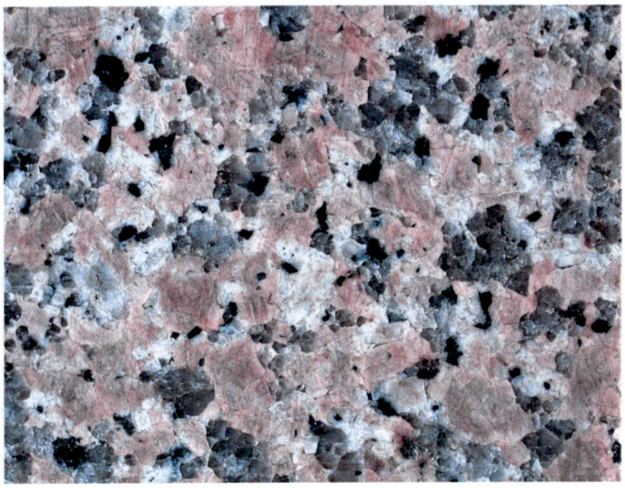

FIGURE 1 Granite with mica, two feldspars, and quartz.

FIGURE 2 Granite with mica, two feldspars, and quartz.

FIGURE 3 The black mica biotite in this granite separated first.

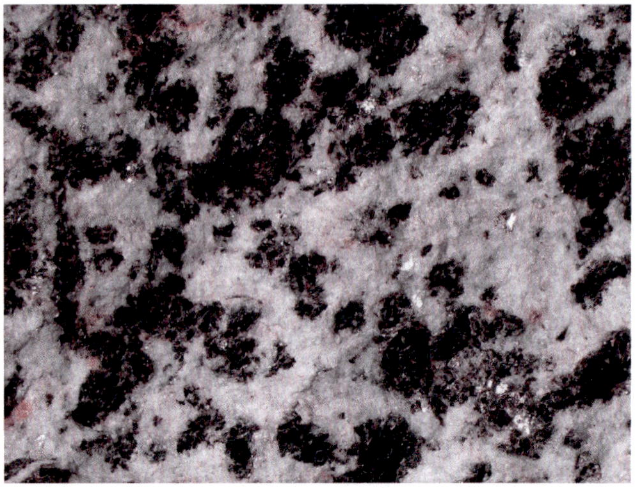

FIGURE 4 The black mica biotite in this granite separated first.

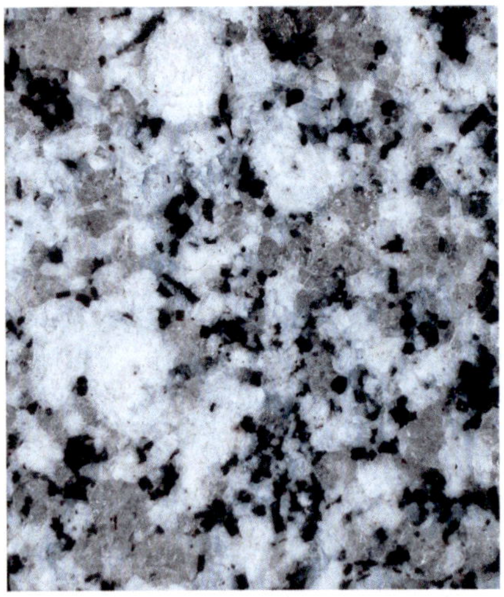

FIGURE 5 The white feldspar in this granite separated first.

FIGURE 6 The white feldspar in this granite separated first.

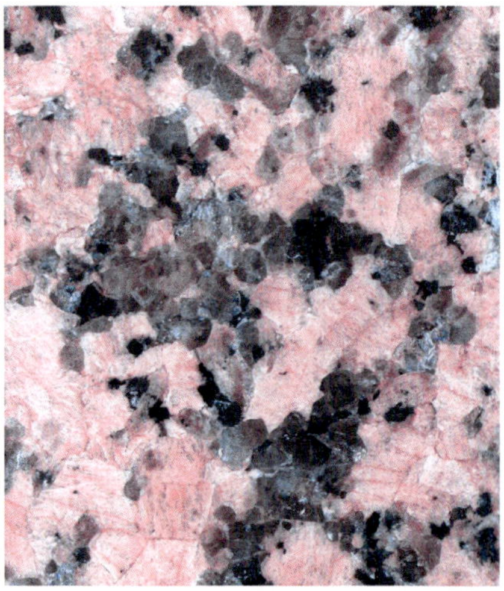

FIGURE 7 The red feldspar in this granite separated first.

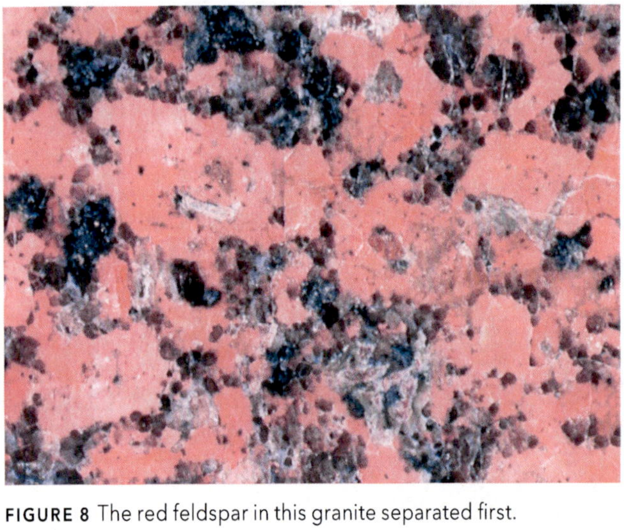

FIGURE 8 The red feldspar in this granite separated first.

FIGURE 9 Gray, translucent quartz filled the remaining gaps in this red feldspar granite.

CHAPTER 2

The Crystalline Rocks *Above the* Granite

Now that we have familiarized ourselves with granite and its four minerals, Figure 10 will guide us through the next topics. It shows a general section through a piece of the Earth's crust with the rocks formed from granite during its evolution. This section is a general picture because each region will have its own variant; some layers may be missing due to subsequent mountain formation, while completely different thicknesses are formed in other places. Below the granite of the continents, there is basalt, which is also under the waters of all the oceans, as indicated on the left side of the section and described in greater detail in Part II.

Above the granite, the first layer of rock is gneiss. The following images show how the four minerals of granite also formed the gneiss in the earliest times of the Earth's history – during the Precambrian (see Geological Timeline on page 108). Mica is found in all gneisses. In Figures 11 to 14, micaceous gneiss is shown with white feldspar and red feldspar. The quartz forms small white grains, and 30% to 40% of every gneiss contains it. The typically lenticular feldspars initially separated as spheres in the hot liquid state and were pressed flatter during hardening and crystallization by the increasing pressure of the additional rock masses. Gneiss can be, on average, 2,000 to 4,000 meters thick.

The gneiss layer is followed by the mica schists, as shown in Figure 15, at around 1,000 meters thick, then the clay schists with quartzite (Figures 16 and 17), which can become several hundred meters thick.

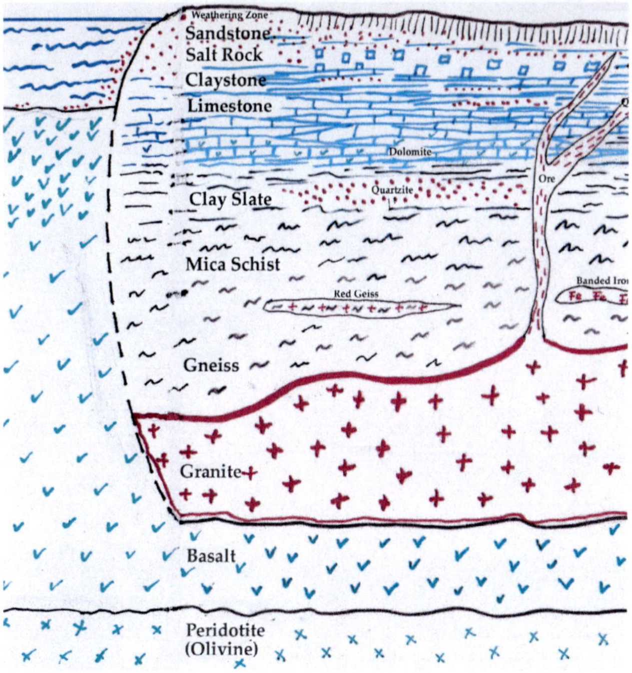

FIGURE 10 Section through the Earth's crust showing generalized rock and mineral layers.

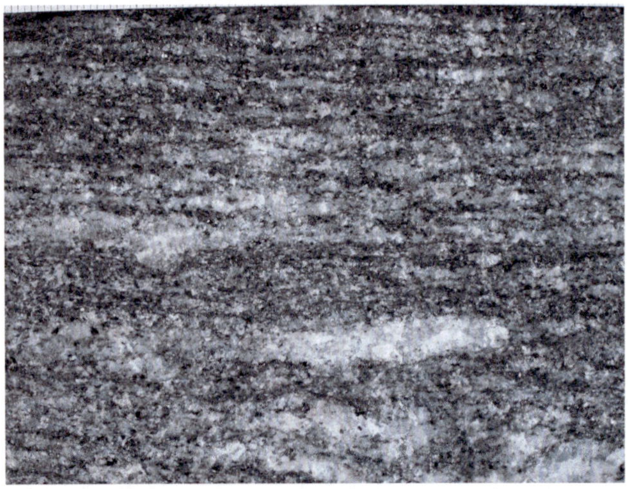

FIGURE 11 Micaceous gneiss with white feldspar.

FIGURE 12 Grey gneiss with folded white feldspar.

FIGURE 13 Micaceous gneiss with red feldspar.

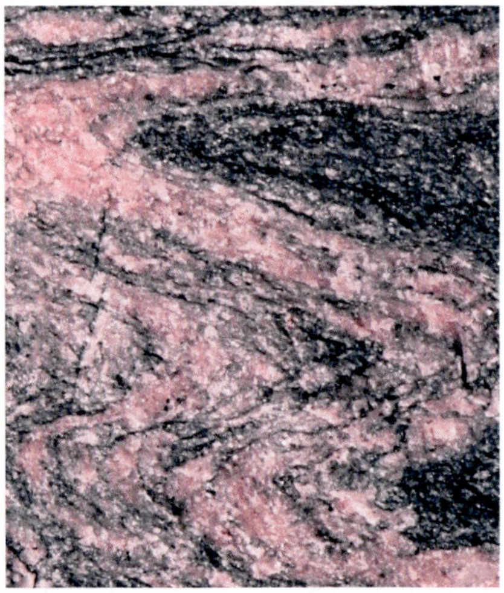

FIGURE 14 Red gneiss in grey gneiss, folded.

FIGURE 15 Mica schist

FIGURE 16 Clay schist with quartzite, folded.

FIGURE 17 Clay schist

The gneisses and schists have been folded many times by later mountain formations during the Earth's history. These crystalline rocks have different thicknesses, depending on the local conditions during the early days of the Earth when they were deposited.

An example of rock layers having variable thicknesses is shown in Figure 18 with a view into the Colorado River Canyon near the bridge southwest of Moab, looking southwest in the left half of the picture towards Zion Canyon and Bryce National Park. The light-colored layers far back in the center of the picture are the same white, crystalline limestones as in the background on the right side and in the front of the picture, on which the viewer stands. They were formed in the Devonian period, i.e., the middle Paleozoic. Beneath them lie brownish quartzites with thin layers of clay schist up to 1,000 meters thick, which form the erosion surfaces carved by the river. The brownish color comes from iron content, which was transformed into limonite, brown iron, under hot and watery conditions. The Colorado River Canyon has only cut into the granite in a few places. This deep river valley was formed from the Mesozoic era onwards, roughly concurrent with the folding of the western Cordillera, when a continental fragment was pushed from west to east under North America. The formation process of every landscape on Earth can be read from the rock layers.

Two processes can be observed in the landscapes. On the one hand, a spatial process was at work shaping the surface of the rock, and on the other hand, a temporal process can be recognized that shaped the Earth's surface in rhythmic steps. This also shows the diverse formations of crystalline rocks, which is difficult to imagine under today's conditions! We may be familiar with this process on the beach, where one wave moves the sand before the next wave flows over it again, and so on. One question remains unanswered: How does the iron come to be so finely distributed among the four granite minerals?

FIGURE 18 Colorado River southwest of Moab, Utah, looking toward Zion Canyon and Bryce National Park.

CHAPTER 3

The Formation *of* Coal

The coals and their related formations have descended from a plant life process into the inorganic through bacterial decomposition. The plant cellulose, chemically a carbon-hydrogen-oxide compound, must be isolated from the air after the plant dies if peat is to form, thus submerged in water. As a result, the carbon is converted from a gaseous state by oxygen and hydrogen into a phase ranging from liquid to solid. Plants cannot survive in deeper water. Peat is formed underwater, and the brown and hard coals are derived from this beginning. These conditions from Earth's history can still be observed in their initial stages today.

Figure 19 shows a peat dig in Finland, where a meter of peat was formed on a wide area over a solid, crystalline subsoil during the Tertiary period. During the Tertiary period of the Cenozoic, brown coals were formed like this but mostly in much thicker layers.

FIGURE 19 Peat dig in Finland.

An example of the extraction process is shown in Figure 20, an aerial view of an open pit brown coal mine east of Berlin. The conveyor bridges in the background are the longest in the world! The two layers of sand above the coal are transported from right to left by the top bridge. Once the coal has been extracted, the pit is backfilled. After a few years, groundwater can create a large lake above the backfilled pit because the coal is no longer present. The entire extraction operation moves to the right, where the coal is present. Under these sands, two excavators are working to extract the lower-quality coal mixed with sand. It is cut at an angle and transported on the lower conveyor bridge to the left side of the pit, where it is dumped and loaded onto trains that run on a long railroad track. It can be sold as loose raw brown coal or pressed into briquettes. These good, sand-free coals are poured as cones onto loading platforms via the two short conveyor bridges at the front, transported on the track to the factory, and processed into brown coal briquettes having almost the same thermal value as hard coal briquettes. The process was invented by the Research Institute of the Freiberg/Saxony Mining Academy and makes a good, inexpensive export product. I visited the mining operation several times in a red visitor vehicle like the one seen here.

Figure 21 shows brown coal with small pieces of amber formed from the resin of coniferous trees. Many wood and plant remains can still be found today in some of Brazil's swamp forests in the Amazon. Other brown coals can also be found in south-eastern Germany near Zittau, on Rügen and south of there, on the Baltic coast, and in the west of Poland, as far as Holland, depending on where the mountain formations brought the brown coals up to a mineable depth. However, this is where the first problem of coal formation begins because, in the Cenozoic, there were no such large and very shallow depressions, about one meter deep, in which plants still grow in the water and form peat bogs from which coal can then be formed!

It is taken for granted in coal geology that the much older hard coals were formed from brown coals.

FIGURE 21 Brown coal with small pieces of amber.

FIGURE 20 Open pit brown coal mine east of Berlin in the District of Cottbus.

Half of the world's hard coals come from the Upper Carboniferous and Permian; smaller deposits come from the Jurassic and the Upper Cretaceous. Deposits worth mining from the Carboniferous occur in Central Europe, around Liverpool, England, in the north of Holland, in Germany in the North Rhine in the Ruhr area, and in the Erzgebirge in Zwickau as far as Krakov in southern Poland. Such shallow depressions, over hundreds of kilometers in which plants can still grow, are unimaginable! However, if you only look at the chemical analyses, the brown coal seems to be turning into hard coal. The carbon content increases in 10% increments from wood at 50%, through peat, brown coal, to hard coal, and on to anthracite, a very solid hard coal, up to pure carbon graphite from the Precambrian at almost 100%. The water and carbon dioxide content decreases accordingly.

There is a *coalification leap* (whatever that means) between brown coals and hard coals because coal research cannot explain it. There are a few other differences. Brown coals react more acidically due to humic acids, while hard coals are alkaline. From peat to brown coal, the plant substances are dehydrated and biochemically decomposed by bacteria. Hard coal is anhydrous and without bacterial processes. For processing purposes, brown coals are classified according to their water content, while hard coals are classified according to their gas content.

Another problem in coal formation is the thickness of the layers. One meter of fresh peat becomes 20 cm of mature peat through dewatering, which over time becomes 5 cm of brown coal. If one assumes a continuous sequence of coal formation, this would yield 2 cm of hard coal. Brown coals are usually up to ten to twenty meters thick. Greater thicknesses of coal seams range from 50 to 100 meters. The thickest brown coal seam on Earth in Morwell/Victoria, Australia, has 320 meters of coal, which would correspond to 6,400 meters of peat. Hard coals are usually only a few meters thick. The world's thickest coal seam in Fushun, China, with 200 meters, would require about 10,000 meters of peat! These are peat thicknesses that cannot be explained by plant growth!

Figure 22 shows a piece of hard coal from the Ruhr area, a famous mining region in western Germany. It consists mainly of two rhythmically

FIGURE 22 Hard coal from the Ruhr area in western Germany.

FIGURE 23 A spherical inclusion of graphite in clay-slate.

alternating layers. Over 90% solid, deep black, structureless, glossy coal (vitrain, vitrite), with white reflective areas, can be seen in the middle layer. Below and above is fine-grained, structureless, less solid matt coal (durain, durite). It rarely contains plant remains such as spores, leaf skins, waxes, and resins, which are only visible under a microscope. During the Carboniferous period (carbon for the coal), the plant world is individualized into evergreen and conifer species and preserved as fossils in the clay shales.

Figure 23 shows a spherical inclusion of completelycrystallized graphite in clay-slate. It is found, for example, in gneisses from the Archaean in South Africa, in the slates of the Paleozoic in Michigan, on Greenland, or in Bohemia near Brno, where it is used for producing pencils. Graphite originated from hard coals, the origin of which has been proven by isotope analyses. Another hypothetical coalification leap is placed between the hard coal and the graphite. Complicated hypotheses are formed because it remains unclear where and how such unimaginable quantities of organic, fat-rich substances could have been formed in the Archaean.

The formation of coal during the entire history of the Earth can only be understood from the gradual death of an airy, watery, albuminous life sphere, as suggested by Rudolf Steiner.

> *It is necessary to turn to extra-human nature and to study the nature of the coal deposits of the Earth, which have advanced more in mineralization. The primary role of carbon in the Earth's process is in connection with oxygen consumption. The Earth's carbon content regulates the oxygen content of its atmospheric environment. The Earth is an organism with the function of respiration, and the carbon content of the Earth has to do with the breathing of the Earth.* (CW 312, March 31, 1920)

CHAPTER 4

The Calcareous-Clayey Sedimentary Rocks

During the living development of the Earth, the plant kingdom was substantially united with the carbonaceous-siliceous and the human-animal kingdom with the oxygenous-calcareous. Thus, the calcareous rocks are formed from the life processes of human beings and animals, which form their skeletal support inside their body or like shells or snails on the outside. Chemically, limestones contain calcium, carbon, and oxygen ($CaCO_3$). The calcium was originally already contained in the white feldspars, the plagioclases of the basalts, and entered the still hot, airy, and watery environment of the Earth through surficial weathering in the Precambrian. Carbon and oxygen are added by the most important living rock-forming organisms, bacteria and algae, especially blue-green algae (cyanobacteria). This is the origin of carbon dioxide (CO_2), which has already been described in the previous chapter on coal formation. The bacterial, plant-animal metabolic activity also created the fine-grained, fossil-free masses of limestone (chalk). Without this metabolic activity, all dead organic matter would have remained unchanged, but the organic remains are processed by bacteria through the process of putrefaction or decay, which converts

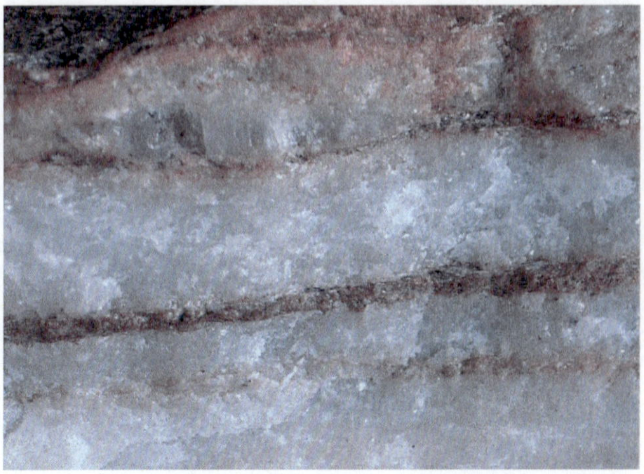

FIGURE 24 Light-colored "dolomite" formed in gneiss.

FIGURE 25 Limestone in rhythmic alternation with clay layers.

them into an inorganic state. The masses of algae multiply on the surface of the water, forming a layer of life that continues to multiply and grow upward. Farther down in the oxygen-free environment of the water, in the absence of air, a digested sludge forms through which the calcium is then bound into a calcareous sludge, which over time densifies into solid limestones and finally crystallizes. The oldest limestones are already found in the Precambrian before animal fossils are found. Figure 24 shows how the lighter-colored "dolomite" was formed in the gneiss. This is followed in the Devonian period of the Paleozoic by limestones in rhythmic alternation with clay layers (Figure 25). These clay minerals originate from the mobilized erosion masses of the granites and crystalline rocks of the Precambrian.

Subsequently, at the beginning of the Mesozoic, additional limestones formed in the Triassic and Jurassic periods in what was then still a warm, watery sphere, as shown in Figure 26. The rhythmically alternating layers can often be seen on the slopes of the valleys in central Germany. The harder limestones (blue) form steep slopes, the soft claystones (green) with their undulating shapes fall off in flat planes, and in between are the harder sandstones (yellow). In addition, sandstone also contains limestone and claystone, limestone contains claystone and sandstone, and claystone contains sandstone and limestone, in each case modifying the individual landscape forms in finer rhythms.

Figure 27 shows a quarry face where very little claystone is embedded in the limestone and forms a complete layer. Similarly, in Figure 28, a lot of claystone was formed, where you can see how the

FIGURE 27 Limestone and a very little claystone, Thuringian quarry.

FIGURE 28 Limestone with a lot of claystone.

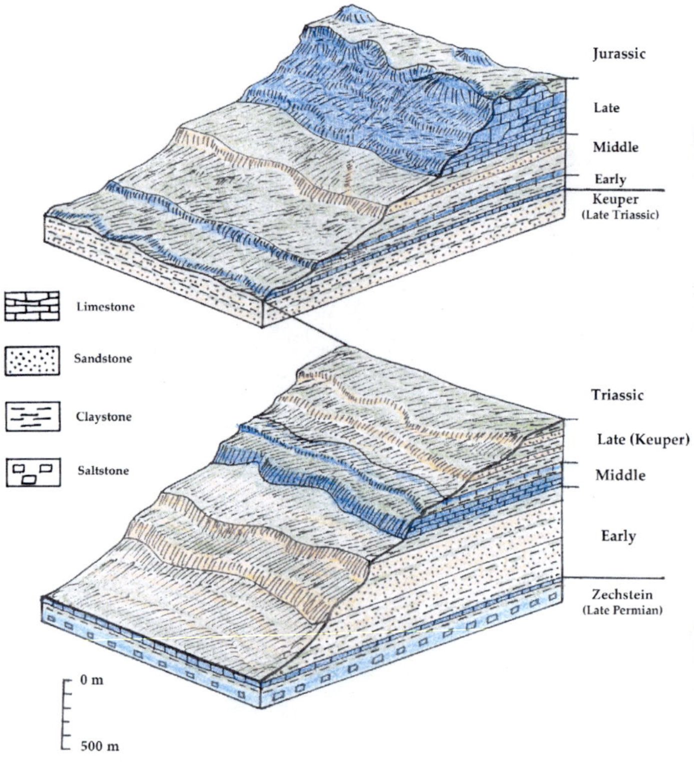

FIGURE 26 Rhythmically changing erosion layers from the Mesozoic in steep steps on the valley sides.

claystone was compressed during the hardening of the limestone. In these limestones, a common fossil is the mussel *Lima striata* on a layered surface (Figure 29).

Different life processes in fine-grained limestones on the steep coast of Rügen, the Baltic Sea island in northern Germany, which rises 70 meters above the beach, are shown in Figure 30. The siliceous content from the shells of lower animals accumulated in these layers during the Cretaceous period and was carried upward with the pore water as it solidified. The silica flocculated as a viscous quartz mass with the organic remains when a certain concentration was reached and then solidified in layers spaced a few meters apart. The animals belonged to shells, sponges, and sea urchins. They can be found as black nodules, sometimes still with the fossilized animals, lying on the beaches of the chalk coast. The organic matter has been decomposed by plant-like, single-celled, millimeter-sized bryozoans and algae, which have colored the chalky white quartz black (Figure 31). This is also shown in Figure 32 by a piece of black and white quartz (top left), and next to it, on the top right, three small, sharp-edged knives fabricated from such a quartz substance. This hard material could also strike fire, so this black or white quartz is called *flint (firestone).* Because of these properties, flints from the beaches of the Baltic Sea were a sought-after export item in Greco-Roman antiquity. And in the lower part of Figure 32 writing chalk can be seen, which was also fabricated on Rügen from the white, fine-grained chalk lime. In earlier times, it was used to write or draw on slateboards in schools up to third grade. The slateboards were fabricated in the south of Thuringia from the fine-layered slate of the Lower Carboniferous. Figure 33 shows a slateboard from my father's workshop, on which the daily work was written. After World War II, I still went to school with a slateboard. A sponge came with it for erasing, and small water containers were inserted in the school desk. Writing chalk became the international

FIGURE 29 The Lima striata mussel.

FIGURE 30 Animals belonging to shells, sponges, and sea urchins can be found as black nodules, sometimes still with the fossilized animals accumulated above the beach along the coast of Rügen.

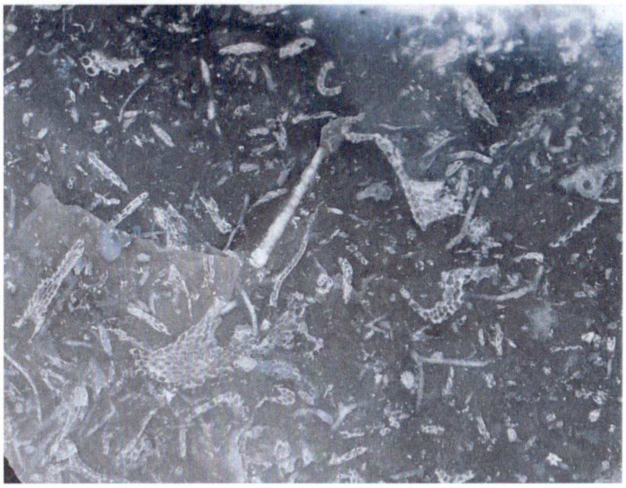

FIGURE 31 Bryozoans and algae turn white quartz black.

name for the last period of the Mesozoic, the *Cretaceous*, meaning *chalk time*!

You can still experience the life activity of bacteria today if you touch a damp, slippery wall in a limestone cave (Figure 34). The groundwater in Europe is around 8° C and reaches the warmer air in a cavity, deposits the dissolved lime in spherical forms, drips down, and evaporates, making the picture seem blurred.

Elsewhere, these globular forming lime processes can still be observed in two subtropical areas of the world, in the Persian Gulf and on the flats of the various spits in eastern Australia, where Figure 35 provides an overview, and Figure 36, a cross-section under and above water. The locals call these limestone formations *living stones* because they seem to grow from the inside to the outside

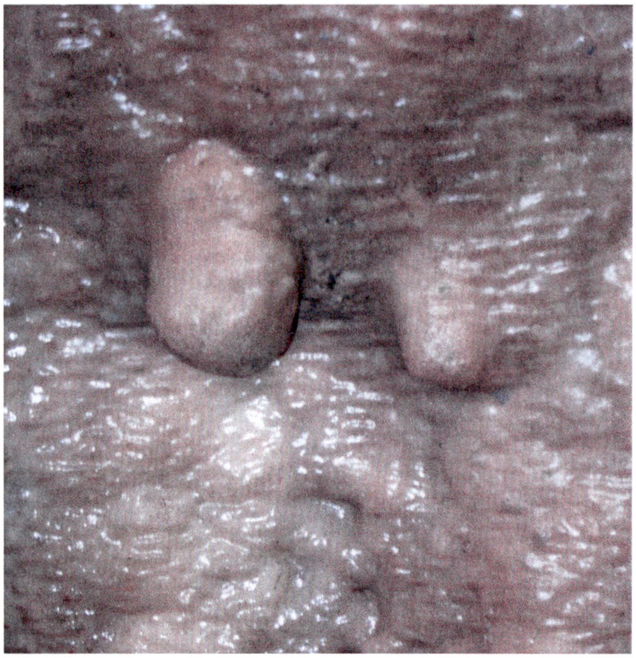

FIGURE 34 Lime, Gypsum anhydrite.

FIGURE 32 Top Left: Chalky white and black quartz. Top Right: three small, sharp-edged flint knives fabricated from quartz substance. Lower: Writing Chalk.

FIGURE 33 Slateboard in my father's workshop.

in the 20°–70° C warm water and deposit calcium carbonate on the upper edge when the tide washes over the forms with new calcium-rich water and some sand. Both photos of this specific type of stromatolite were taken at low tide. These bacterially formed limestone layers, as described here, are only a few hundred meters thick on the American and European continental surfaces, but they are several thousand meters thick in the fracture zones between the mountain blocks, for example, in the Alps, in the Himalayas, in the Altai, in California, or in the Andes. Gypsum and salt rocks are often embedded in these limestones, which are considered in the next chapter.

FIGURE 35 Globular forming lime processes formed these "living stones" in eastern Australia.

FIGURE 36 Globular forming lime processes formed these "living stones."

CHAPTER 5

The Gypsum *and* Salt Rocks

Gypsum rocks and most salt deposits were formed in the Permian period at the end of the Paleozoic. In a chemical analysis, gypsum is a calcium sulfate dihydrate ($CaSO_4{\cdot}2H_2O$) and similar to calcium carbonate ($CaCO_3$). It is also a step in the development of the white feldspar, a calcium-sodium sulfate ($Ca,Na\ Si_3\text{-}O_8$), whereby the sodium leads to rock salt ($NaCl$). Sulfur, oxygen, and chlorine are added by the life processes of bacteria and algae that lived in the warm, watery atmosphere of the Earth's environment at that time. Salt-absorbing bacteria live optimally in a thirty-percent rock salt solution, thus, close to the salt's saturation limit. Initially, gypsum was formed from anhydrous anhydrite, like the samples shown in Figures 37 and 38. Figure 37 shows an example of a dark limestone and gray anhydrite already merging into the lighter patches of gypsum. Figure 38 shows these three rocks again, in which the volume of the white gypsum has increased by 60% due to water absorption. As a result, it folded and pushed the calcareous anhydrite layers upward! When the

FIGURE 37 Calcite, gypsum, and hydrate.

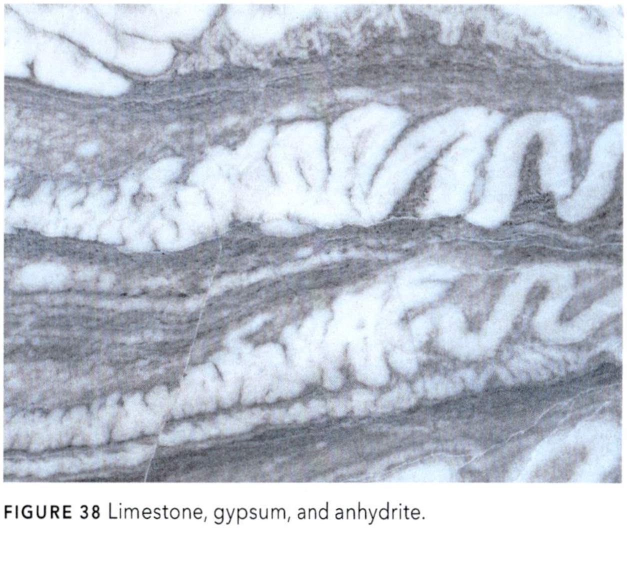

FIGURE 38 Limestone, gypsum, and anhydrite.

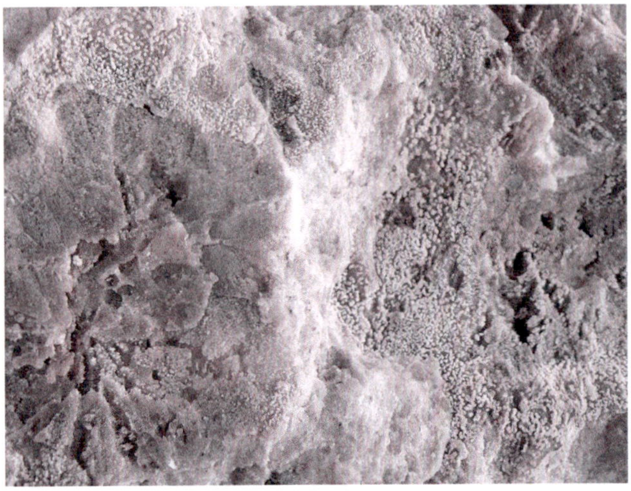

FIGURE 39 A layered surface of gypsum in a cavity.

FIGURE 40 A landscape folded and pushed upwards by water-saturated gypsum.

gypsum begins to crystallize in the water-filled cavities, various crystal forms can develop and become transparent (Figure 39). When the gypsum layers come to the surface, they are easily eroded, and a landscape is created like this one near Lausanne on Lake Geneva in Switzerland (Figure 40).

Figure 41 shows a cross-section through the sediments of the Earth's crust that formed on the bedrock below. It shows a *salt dome,* a characteristic feature of salt rocks. Compared to the other sediments, rock salt is softer and was pushed upward by pressure where faults formed underground. This feature is particularly evident in the Zechstein in the late Permian period at the end of the Paleozoic (blue with dashes). Above this comes the Bunter Sandstone (Buntsandstein) in the Lower Triassic period of the Mesozoic (yellow), the Muschelkalk (blue), and the Keuper marls and clays with some gypsum in the Upper Triassic (light yellow with blue dashes). The next three layers mark the Jurassic period: the Lias (light yellow with yellow dots), Dogger (similar, darker yellow with dots), and the Malm with sandstones (blue with yellow dots). Then come the sandstones of the Cretaceous period (yellow) and the Cenozoic sediments (white) that formed in the Tertiary period concurrent with the high mountains of the Alps in central and northern Germany. This last activity created the faults mentioned above, where the volcanoes on the left side (blue) and the salt domes cut through the layers of the Mesozoic. These changes brought the salt into the groundwater zone, where underground *salt lakes* formed.

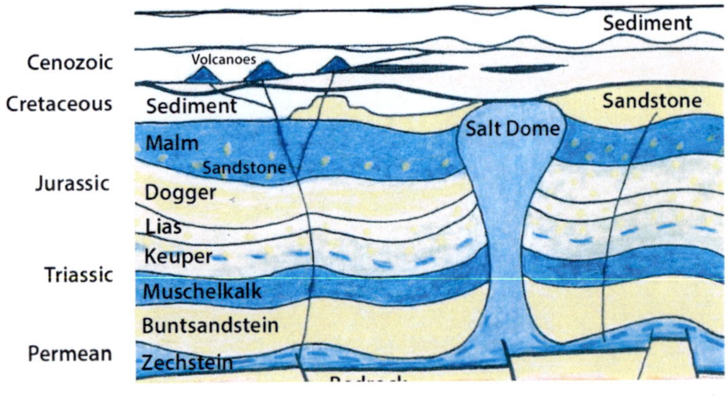

FIGURE 41 Cross-section through the sediments of the Earth's crust in Europe, with formation names from German lithostratigraphy.

From around the 12th century, the saltwater from salt lakes was extracted through boreholes and fed into large shallow basins at *salt boiling works,* which were heated from below so the water evaporated and dried *table salt* could be extracted. For example, in Lüneburg, in northern Germany, the *Hanseatic League* was formed, and its ships carried salt to Scandinavia for curing fish, which was a source of rich revenues. Culture flourished in the Hanseatic cities during the Gothic period; richly decorated churches

FIGURE 42 World's largest salt works in Bad Dürrenberg.

FIGURE 43 Gypsum crystallized around pine branches in the salt works.

and merchant houses were built, as in Hamburg, Kiel, Rostock, or inland in Frankfurt/Main, Erfurt, or around Halle/Saale, where salt was also mined and extracted by boiling; or around Salzburg and Hallein, on the northern edge of the Alps.

Another method of extracting salt was developed in central Germany during the 18th century with the construction of salt works. Figure 42 shows the world's largest salt works still operating in Bad Dürrenberg. Pine branches were staked in a wooden frame (light brown in the figure). With elaborate wind power technology, salt water was pumped up to the top to flow evenly over the pine branches. The gypsum in the salt water settles on the branches, increasing the water's salt content from 7% to 22%. Today, the system is used by lung patients who can walk on the walkway below, inhaling the vapors. After certain periods, the loose packets of branches are so filled with gypsum that they must be replaced, two pieces of which are shown in Figure 43. You can see how the gypsum has crystallized around the branches, radiating outward. It cannot be used any further because the pine branches are in it. Nowadays, gypsum is used to make sulphuric acid (H_2SO_4). To prepare gypsum for use as stucco plaster, it must be heated to 120-130° C, dehydrating most of it.

FIGURE 44 Evaporation area in Australia, middle foreground.

FIGURE 45 Solid lake surface with a salt crust that contracted through crystallization.

FIGURE 46 Salt evaporation ponds, Lanzarote, Canary Islands.

FIGURE 47 Potash mine in Sondershausen.

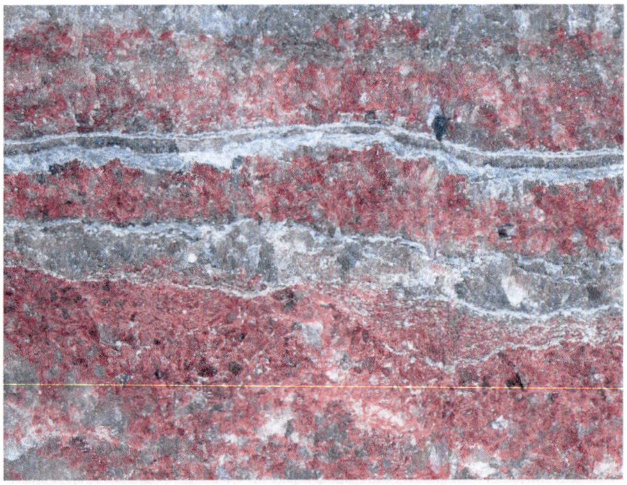

FIGURE 48 Potassium salt from the mine at Sondershausen, Thuringia.

Figures 44 to 46 show how gypsum naturally separates from salt water in warm countries near the equator. Figure 44 shows a wide lake in Australia with a surface firmly crystallized in the darker area where my friend stands. The dilapidated wood structure is from an old jetty. Farther out, the open water surface can be seen in the whiter strip. Figure 45 shows the solid surface of this lake, which has contracted and cracked during crystallization. A third way to extract salt is through *salt evaporation ponds* (Figure 46) in Lanzarote of the Canary Islands. Sea water is pumped into the ponds through pumps in the three towers at the center of the picture. The heat of the Sun and the constant wind dry the salt; then, the pond areas are scraped out, and the salt is collected in white piles for transport.

A further development of salt rock at the end of the Paleozoic is potassium salts, which have been increasingly used for artificial fertilization since the end of the 19th century. This use has resulted in insects no longer fertilizing fields, as most species are becoming extinct. The most important potassium salts are sylvan (KCl, 62% K_2O), carnallite (KCl with Mg, 17% K_2O), and other salts with amounts of K_2O in between. In addition, small amounts of red iron hematite can give the potassium salts a red color. These chemical elements point to substances originating in the early life sphere, including potassium, magnesium, red iron, and black mica biotite; oxygen and chlorine point to bacterial life.

Figure 47 shows a mining area in the potash mine at Sondershausen, Thuringia. At the seam on the left, the author taps out a sample of potassium salt, a closeup of which can be seen in Figure 48. The yellow lighting outshines the red and white rhythmic layering of white rock salt and red potassium salt. In the center-left is a mining machine scraping the path through the cavity. The remainder of the path is still lower and has yet to be excavated. The potassium salt is loaded onto special trucks that are hauled to the surface from this depth of 600 meters.

Figure 48 shows the salt sample that I tapped out of the wall. The red potassium salt alternates in layers with white rock salt. East Germany was the world's second-largest potassium salt producer after Canada! A dedicated loading port was built in Wismar on the Baltic Sea. After reunification in 1989, all potassium salt mines were closed, and around 20,000 miners and transporters were made redundant!

Rudolf Steiner describes:

> *It is especially characteristic for a human in the present phase of Earth's evolution to share in the formative process of the mineral. If we require a specific term for this process, let it be derived from the most characteristic feature, namely the process of de-salification. Within our human organism, there continually takes place a process of de-salting, a tendency to change salt formation into its opposite; on this, our being human rests, and above all, our human thinking, which goes beyond the animal.* (CW 312, April 4, 1920)

CHAPTER 6

The Sandstones, Claystones, Gravels, *and* Sands

Sandy-clayey rocks are formed from solid crystalline rocks that weather on their surfaces, then are eroded by the swirling air-water sphere and washed into large depressions throughout Earth's history. There they solidified in layers with water diffusing upward from below. Figures 49 to 51 show sands deposited in the early Paleozoic era and then crystallized into solid quartzite. In Figure 49, the light-colored layers of quartz sands alternate with darker clay layers. The coloring comes from iron ore that oxidized in water to hematite, a red iron ore. Following the *angled cross-stratification*, you can see how the already firmer layers were cut off again and covered by new layers. Due to strong currents, such processes occurred at the bottom of the extensive depressions. This can also be seen in Figure 50, a layer in Australia on which the fine-grained sand was moved from the bottom right to the top left and then hardened before the layer was overlaid by the next. Figure 51 shows another quartzitic layer sequence of light quartz sands and darker clay sands with small black mica grains.

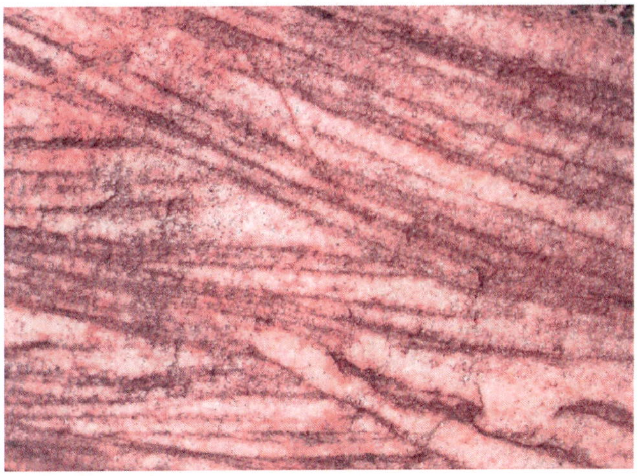

FIGURE 49 Quartzite with cross-bedding.

FIGURE 50 Quartzite in distinct layers, Australia.

FIGURE 51 Quartzite with lighter quartz sands and darker clay. sands with small black mica grains.

FIGURE 52 From bottom to top: coal sands, clay sands, and yellowish sands in the groundwater zone.

This brings us to the sands and gravels of the Cenozoic era. In Chapter 3 on the formation of coal, an open-pit brown coal mine is shown in Figure 20, where loose sands deposited by the Elbe River in the Tertiary period before the Ore Mountains (Erzgebirge) were uplifted, can be seen on the top right and through which the river then cut its way. These sands, transported on the long conveyor bridge to the left, can be seen closely in Figure 52. At the bottom are the coal sands, boulder horizons, and fine quartz sands above them, some with layers of clay sand containing plant remains. At the very top, the sands in the groundwater area are colored yellowish by brown iron. *Shifting sand dunes* in Namibia were formed where the sand had not yet solidified, as seen in Figure 53. They consist of fine sand where every grain is covered with brown iron. But these dunes are only shifting slightly, as evidenced by the grasses growing to the right.

Figure 54 shows the volcanic landscape on the Canary Island of Lanzarote. When the volcanoes erupted a few hundred years ago, the lava flew several hundred meters into the air, fell in solidified chunks and ash, and also formed flat dunes.

This concludes the topics on the development of granitic minerals and leads to the basaltic rocks, whose minerals have always shaped the granitic rocks through their substances and colors.

FIGURE 53 Shifting sand dunes, Namibia.

FIGURE 54 The volcanic landscape of the Canary Island Lanzarote.

PART II

The Formation *of the* Earth

CHAPTER 7

The Separation *of the* Planets *in the* Solar System

The beginnings of planetary development are shown schematically in Figure 55. This figure shows the separation of the planets contracting into their own spheres of activity until the Earth and Moon separated in the middle of the Lemurian epoch. Rudolf Steiner only drew the circles designating the planets on his blackboard drawing (CW 102, February 15, 1908). I added the words, colors, spheres, planetoid belts, the basaltic (green), and the Sun's radiation. The planetary spheres gradually contracted from the Polarian age before the Archean eon. The planetary bodies became visible at the outer limits of their orbits. Uranus has but slight connection with our earthly existence. Colonies of beings of the highest spiritual hierarchies work within the Sun's sphere, whereby the planetary bodies are separated into their own orbits to create for the human being with all its constituent parts and the living kingdoms of nature, a dwelling place on the Earth. First, the physical effects will be described and then the further developments.

The rings of Saturn were formed by basaltic dust and micrometeorites smaller than 1 mm, as the Cassini probe determined in the year 2000. Jupiter, the largest planet in the solar system, physically consists of gases that orbit the planet's body in huge rings at breakneck speed. Like Saturn, it has rings held in place by its gravitational and magnetic fields and has attracted numerous cosmic bodies. That Saturn and Jupiter can be seen as planets in the night sky is

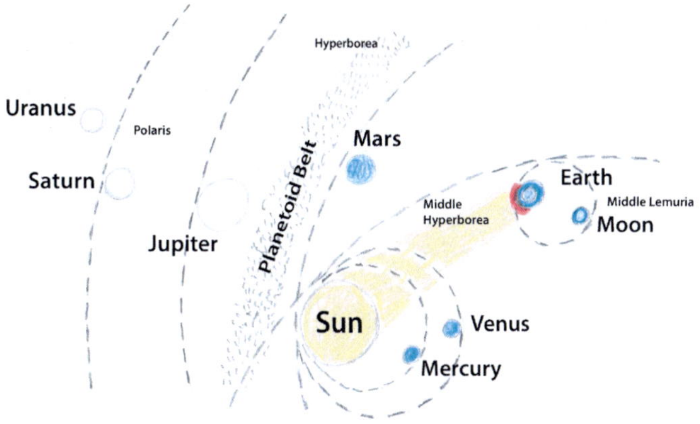

FIGURE 55 The separation of the planets contracting into their own spheres of activity until the Earth and Moon separated in the middle of the Lemurian epoch. The sizes and distances cannot be represented to scale in a drawing.

mainly because of water ice, with a small proportion of hydrogen, helium, sulfur, and ammonia.

In the Hyperborean epoch the planetary bodies condensed into a luminous, heated air state. The planets above the Sun – Saturn, Jupiter, and Mars – separated within the solar system as a physical expression of the development of spiritual beings and human beings. Rudolf Steiner described in detail several times how the substance of the planets remained behind at the outer edge of their respective orbits as they contracted (CW 102, February 15, 1908). In the middle of the Hyperborean epoch, the planets below the Sun – Venus, Mercury, and the Earth-Moon – were still within the Sun-sphere. The etheric was imprinted on the Earth from the planetary spheres on the side facing the Sun (red). In this way, a warm-airy sphere of life was created on the continental side of the Earth. The human being lived in a still barely differentiated air-warmth body in this life sphere; it further developed, and then the nature kingdoms individually came about, which can be understood as the emergence of life.

However, this development causes destruction, about which Rudolf Steiner says:

> *Scattered between the orbits of Jupiter and Mars are a multitude of small planetoids. This is the battlefield wreckage of the war in heaven that was waged between two points in cosmic time when our solar system was contracting, first to Jupiter and then to Mars.* (CW 110, April 14, 1909)

The meteorites and the regularly recurring comets appearing in the Earth's circumference originate from this planetoid belt and reach every part of the Earth's surface. This is how the iron-nickel meteorites fall to Earth, especially in the autumn. Figures 56 and 57 show how the basaltic minerals from the planetoid belt were crushed: The light gray and the brownish feldspar, the greenish olivine, and the white quartz were crushed before they were enclosed in black magnetite.

Meteorites are particularly abundant in the Antarctic, and to some extent also in Greenland, where glaciers are melting all around the edges, leaving the meteorites on top where they can be collected by the hundreds! They fall in similar quantities in other areas of the world, only they are rarely recognized. It was different in Kiel in northern Germany, where a meteorite fragment smashed a garage roof! The meteorites are attracted by the Earth and fly through the atmosphere at a temperature of around 2,000° C and a speed of 20 km per second, i.e., the ionosphere from an altitude of around 800 km and then the stratosphere from an altitude of 65 km,

FIGURE 56 The light gray and brownish feldspar, greenish olivine, and white quartz were crushed before they were enclosed in the black magnetite.

FIGURE 57 The light gray and brownish feldspar, greenish olivine, and white quartz were crushed before they were enclosed in the black magnetite.

where they light up as *shooting stars* as they burn up. Only a small amount of this substance reaches the Earth, but the impact pressure can cause deep craters. To the east of Stuttgart, for example, the limestone from the Jurassic period was penetrated some 600 meters deep down to the granite, and a basin with a diameter of 1,500 meters was created, in the middle of which Nördlingen is located today. However, no meteoric iron was found because the iron substances probably vaporized upon impact. Other impact sinks created by meteorites have also become famous, such as the Arizona meteorite crater, with a diameter of over 1 km and a depth of 170 meters, in which even larger chunks of meteoric iron were found underneath.

Figure 58 shows the Hoba meteorite in Namibia, where I could scrape out small pieces from under the block and strike them at home. Figure 59 shows the minerals. The dark gray magnetite, with black, roundish grains, makes up the groundmass and contains 6-10 % nickel, which contains white feldspar and dark green olivine. Two clear phenomena can prove that a piece of iron ore comes from a meteorite. One is the cavities on the surface, which are dislodged by the Earth's watery air layer on approach. The other is the Widmannstätten Figure (Figure 60), which shows the black magnetite and the light-colored 10 to 30% nickel ore. This octahedral structure is formed at several hundred degrees and cannot be reproduced in the laboratory. If you have a questionable piece, a tombstone cutter can grind it for you. Referring to Figure 55 above, moving from left to right, the Sun's direct sphere of activity begins after the planetoid belt with the basaltic solid planets Mars, Venus, Mercury, and the Moon in the Earth's circumference. It is on the Earth where the living kingdoms of nature could develop.

Figure 61 is a photograph taken during a waxing Moon with the Sun illuminating the surface from the right, creating only light and shadow. You can see brighter areas interspersed with dark round spots from the Earth, especially at full Moon. They were called *terra* and *mare* regions in the 19th century because it was thought that there were landscapes there, like on Earth with continents and seas. But the Moon has no atmosphere like the Earth. The figure shows the northern edge of *Mare Imbrium* at the front and the *lunar Alps*, *Mare Frigoris*, behind it. The relative age of the craters can be recognized by their morphology: the oldest are the indistinct trench craters of the Terrae; after that came the impacts on the Mares, caused by meteorites hitting them unchecked. The *alpine valley* in the foreground on the right was also torn open in the process, which shows in its straightness how far the crust had already hardened. The mare impact craters were flooded by

FIGURE 58 Hoba meteorite in Namibia.

FIGURE 59 The minerals of the Hoba meteorite.

FIGURE 60 A Widmannstätten pattern, showing a meteoric form.

FIGURE 61 Waxing Moon with the Sun illuminating the surface from the right (Newell, 1969).

basalt, which also filled some of the terrae's marginal craters, such as the large crater at the front left, which solidified very quickly because the temperatures on the side where the Sun shines are around 100 degrees plus centigrade and 160 degrees minus on the dark side. After that, only relatively few impact craters of a few cm to 120 km in diameter were formed in the upper Tertiary period, but no volcanic craters, which mainly date from when the Moon was formed. This is almost the only type of landscape on the Moon and only the trenches at that. There are four to ten times as many craters in the terra areas as in the mare areas, where lunar expeditions first landed because the ground is smoother. A surface structure similar to that of the Moon has been recognized on the surface of Mars. *Regolith*, which consists of broken basalt debris mixed with brownish-greenish rock glass that solidified very quickly from molten basalt or in fragments, is widespread everywhere under the fine dust. After the first rocks were brought to Earth from the Moon in 1969, the approximately 400 kg were analyzed at several locations using all the methods of modern laboratory technology, making these rocks the best-studied substance in the world. It was confirmed that the minerals are similar to terrestrial basalts and peridotites, such as magnetite with titanite, white feldspar, and quartz.

CHAPTER 8

The Planets *and the* Ore Minerals

The seven primary metals of the ore minerals can be summarized in a heptagon (Figure 62). Their formation is best understood through the separation of the planets in the solar system. This illustration can be used to describe the various relationships of the planets to the development of human beings. Rudolf Steiner states:

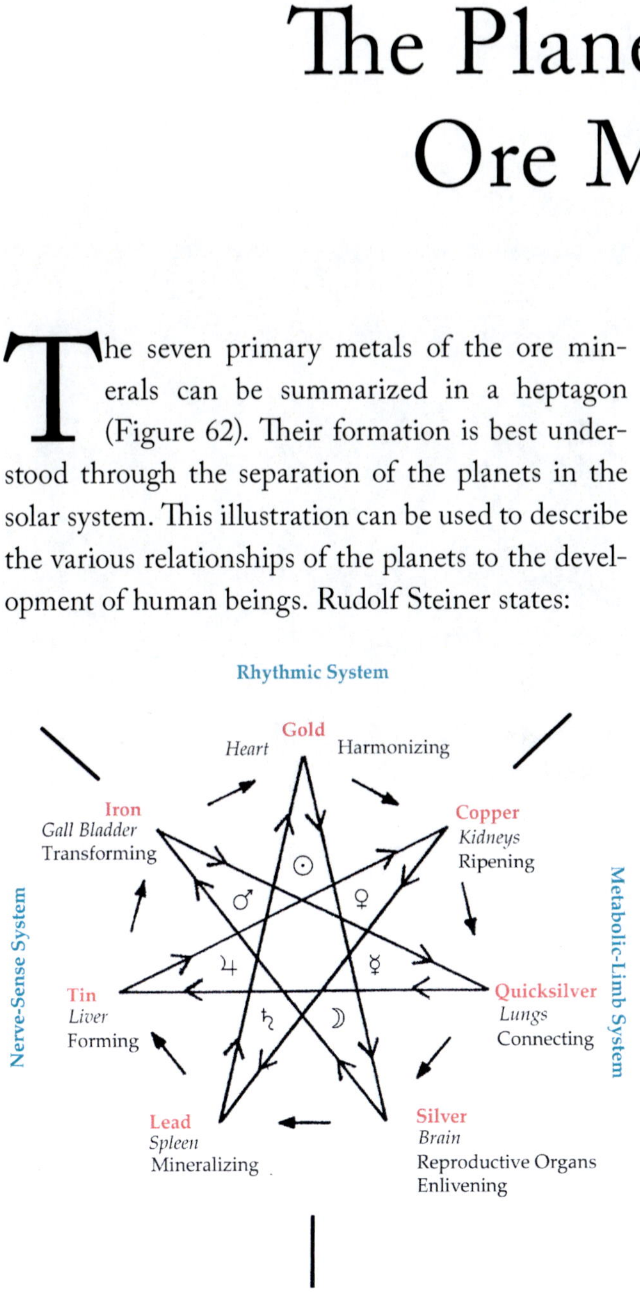

FIGURE 62 A heptagon with the seven principal metallic elements of the Earth's ore minerals (red) and their relation to the evolution of human beings (italics) and the planets (symbols).

In ancient times, there were occult symbols for this. A heptagonal sign is the symbol of the good Sun spirit. The seven corners symbolize the seven planets. The forces are all intertwined and surrounded by a line. They connect everything together." (CW 96, April 27, 1907) Later, he adds: "In a natural mineral system, the relationships of the metals to each other would have to be taken into account, and we would see that in such a system, the lead compounds, the lead itself, lie at one pole, that silver lies at the other pole, that in the middle, for example, is the aurum, the gold, and the others are then arranged accordingly. (CW 313, April 17, 1921)

The outer circle in Figure 62 identifies the threefold formation of the human organization (green). We must imagine the zodiac surrounding the heptagon, from which the highest spirits have brought about the development of the human being. Going further toward the center of the picture, the individual black arrows in the inner circle point to the seven main metals (red) that formed the ore minerals. The right side shows the precious metals; the left side shows the base metals. Under the names of the metals, the corresponding organs are indicated in italics, with the associated organ activity below that in black type. The arrows on the lines in the heptagon, which encloses the planetary symbols, show the developmental sequence of humanity in its seven states of consciousness, as described by Rudolf Steiner from Old Saturn through the states of the Old Sun, Old Moon, Mars, and Mercury as Earth, Jupiter, and Venus. If we start with Sunday and follow the lines with embedded arrowheads, the sequence corresponds to the days of the week up to Saturday: gold, silver, iron, mercury, tin, copper, and lead. This is presented by Rudolf Steiner in lectures, from which I have compiled this heptagon and adapted it accordingly. (CW 96, April 27, 1907; CW 104, June 27, 1908; CW 119, March 22, 1910; CW 313, April 17, 1921; CW 233a, April 21, 1924; CW 279, July 7, 1924).

The following seven figures show the metal ores featured in the heptagon.

FIGURE 63 Gold

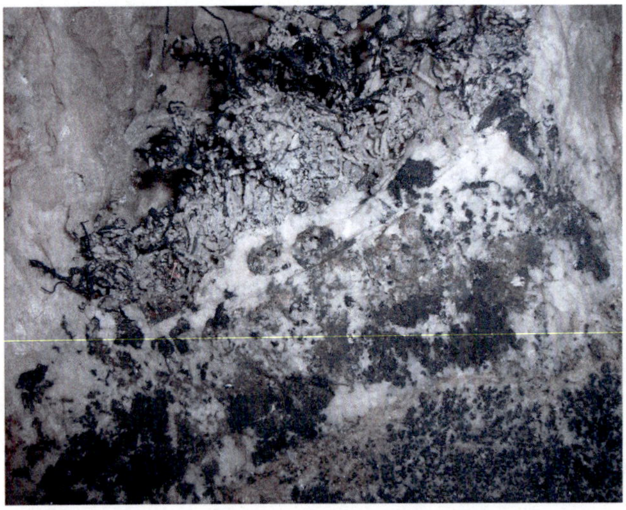

FIGURE 64 Silver

FIGURE 65 Iron

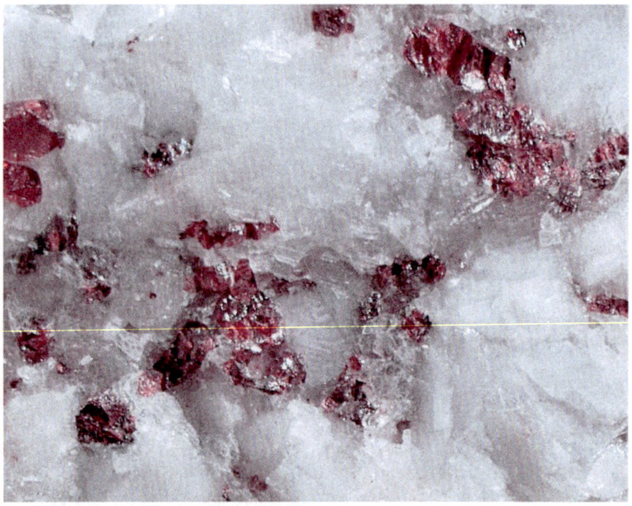

FIGURE 66 Quicksilver, Mercury

FIGURE 67 Tin

FIGURE 68 Copper

FIGURE 69 Lead

CHAPTER9

The Rocks and Minerals *of the* Basalts

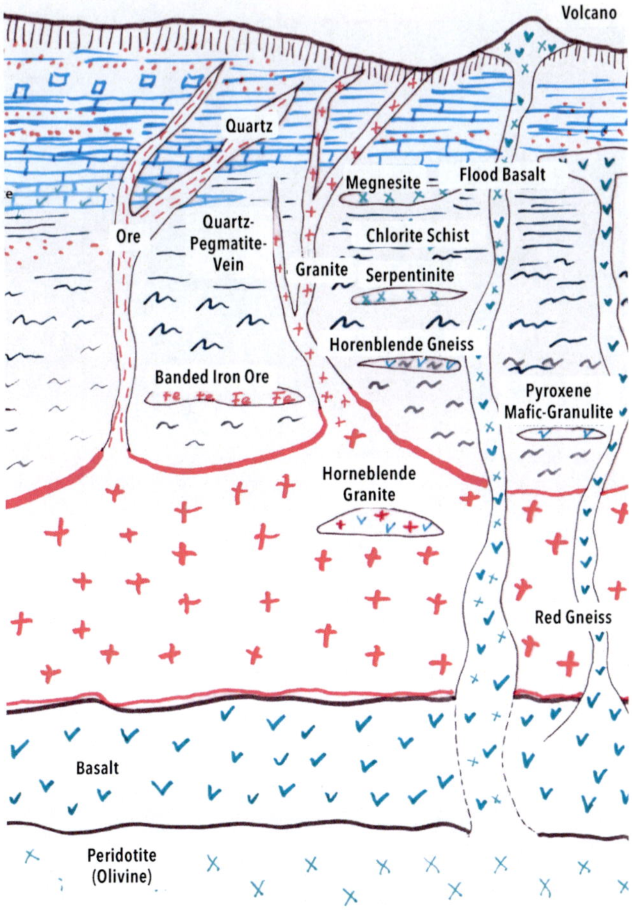

FIGURE 70 Section of the Earth's Crust showing crystalline rocks and basaltic vents.

All cosmic bodies are made of basaltic minerals, as shown in Chapter 7, Figure 55. On Earth, they still form the bottom of the oceans and the foundation of the continents. Where mountain movements create deep cracks, the red-hot basaltic masses can rise high enough to penetrate the Earth's surface.

The basalt vent to the far right of Figure 70 has penetrated all Earth layers originating from the granite above the basalt. Below the granitic are the basalt layers and below this, the olivine basalt, called *peridotite.* They form the upper mantle, about 30 kilometers thick, or about as thick as a paper layer on a globe. In the lower mantle, the iron-nickel zone is presumed to cause the Earth's magnetism. This is how deep earthquake waves go. The Earth has basalt and peridotite in common with the solid planets and all other planetary bodies. To the right, peridotite has also cut through the rock layers via fissures. In addition, during earlier geological periods, flood basalts and banded iron ores reached the Earth's surface through volcanoes. Tertiary volcanoes also brought flood basalts up to the Earth's surface.

In Iceland, a volcanic fissure breaks through a glacier that flows from the top left to the bottom

FIGURE 71 A volcanic fissure breaks through a glacier in Iceland (Schutzbach 1985).

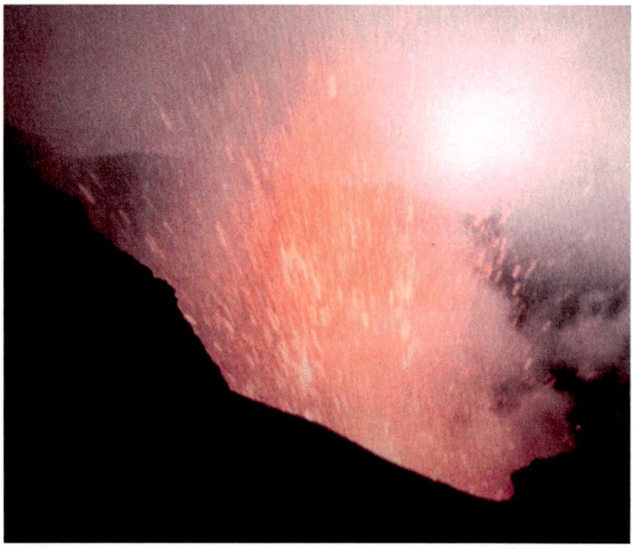

FIGURE 72 Stromboli volcano erupting.

right in Figure 71. I also visited volcanic craters in southern Italy, *Mount Vesuvius*, then in its dormant phase, and *Mount Etna*, which had just blocked the access roads at the time. I was also on the summit of *Stromboli*, which is over 900 meters high (Figure 72). There are two vents in the crater up there; one erupts regularly every 20 minutes or so, and the other explodes independently, suddenly, and strongly. This volcanic mountain was formed in pre-Christian times and has been getting higher and higher ever since. This happens because water at the bottom of the sea enters a vent, where it is heated to around 2,000° C, and then the basaltic material is transported upward, similar to a geyser.

The next two pictures show basalt deposited during the Tertiary period in Iceland. Figure 73 shows the rhythmic alternation between basalt outpourings that formed columns and ejected ashes that ascended high into the air before falling to Earth as a loose material. This deposition of softer material created the flatter slope and the escarpment above in the background landscape. These rhythms can also be observed in the landscape of other volcanic areas. Basalt in Armenia (Figure 74) shows how the flowing, red-hot mass comes to the cold surface of the

FIGURE 73 Columnar Basalt and landscape shaped by loose basaltic material and ejected ashes in Iceland.

FIGURE 74 Basalt columns in Armenia.

FIGURE 75 Lava poured onto a colder surface, taking on rounded, bulging forms.

Earth, contracts, cools from the outer surface inward and tends to form columns under the conditions present there. In Figure 75, lava forms rounded, globular layers when it flows onto a cold surface and bulges before the mass is solidified.

The rocks in the next three images contain about 30% greenish olivine, known as *peridotite*; it is marked with green crosses in the Earth's crust section shown above in Figure 70. Figure 76 shows the three primary minerals in the basalt: black magnetite, light gray to brownish feldspar, and white quartz. In Figure 77, the green olivine and the white feldspar are clearly crystallized in black magnetite; additional ore minerals from the planet's orbit may also be present. The magnetite comes from the basalt. The olivine only comes up to the Earth's surface when brought along with the basalt (Figure 78). The minerals also clearly show that the Earth with the granitic continents of such thickness could not have originated from the basalts, and certainly not the living nature kingdoms with human beings. This leaves many questions unanswered, which will be resolved in Part III.

FIGURE 76 The three main minerals of basalt: black magnetite, light grey and brownish feldspar, and white quartz.

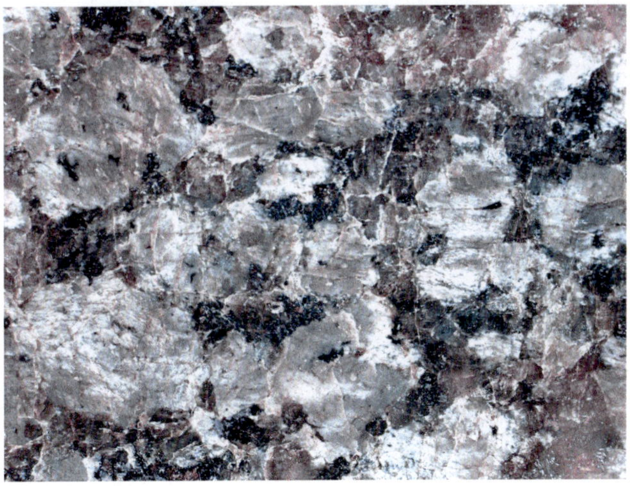

FIGURE 77 The green olivine and white feldspar are clearly crystallized in the black magnetite.

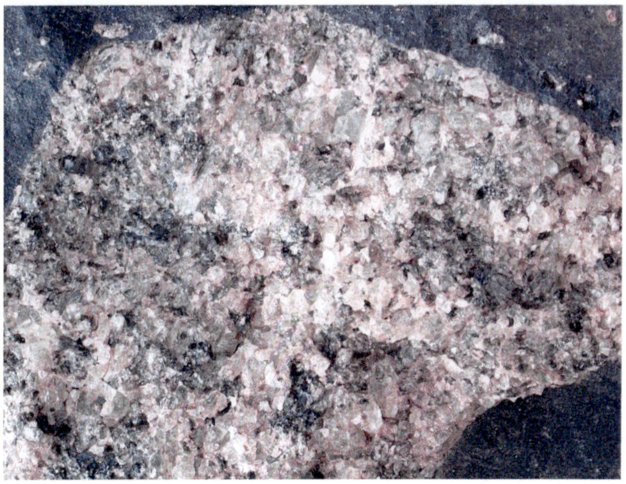

FIGURE 78 Peridotite in basalt.

CHAPTER 10

The Formation *of the* Continents *from the* Life Sphere

In the section of the Earth's crust in Figure 79, on the far right, you can see how the basalt vents of the flood basalts and volcanoes have penetrated all the layers of the Earth that have developed from the granite. On average, the Earth's continental crust is about 15 to 20 km thick! The Earth's diameter is about 12,800 km, so the crust is wafer-thin in comparison! The granite layer is around 4,000 to

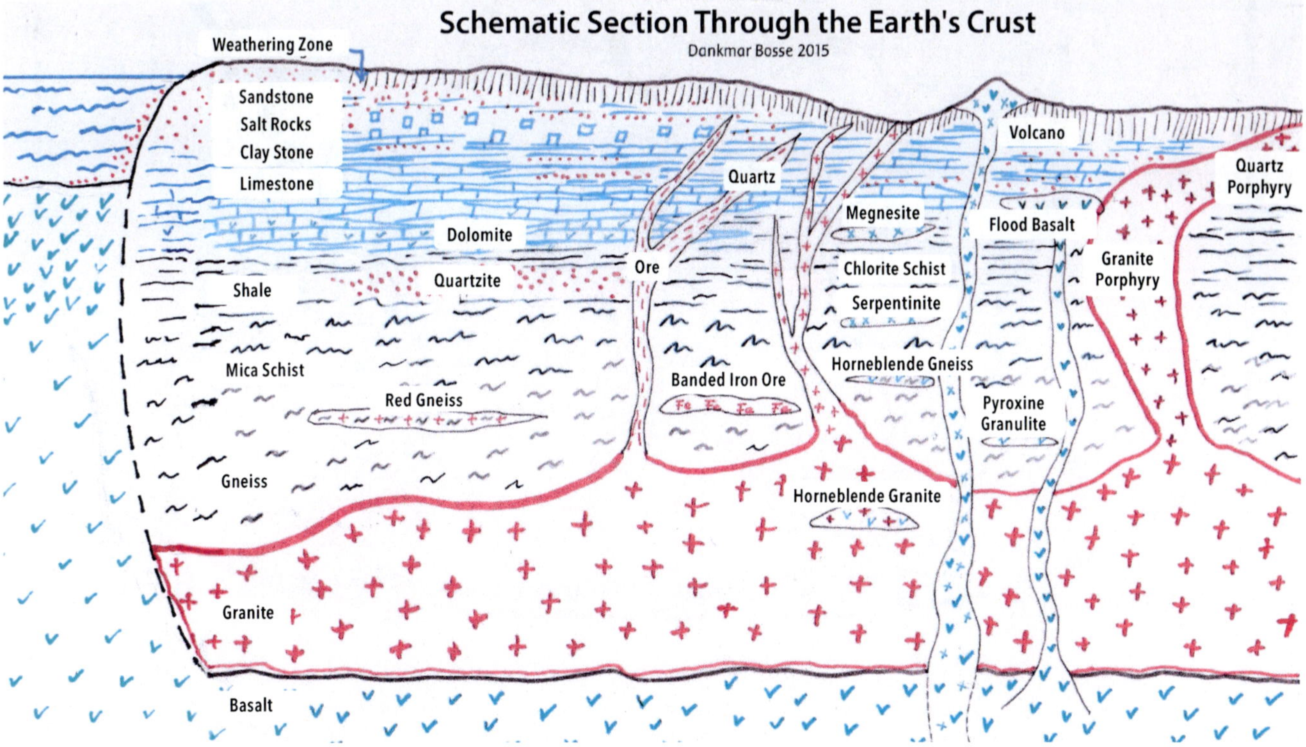

FIGURE 79 General rock sequence in the continents and under the ocean.

6,000 meters thick; the gneiss above it can be 2,000 to 4,000 meters thick, the mica-schist around 1,000 meters, then the clay schists with the quartzite, and the limestones with the mudstones are each several hundred meters thick. Thus, the rock-forming processes have decreased as the Earth ages.

Figure 80 (below) shows the structure of the large supercontinent Pangaea when all continental areas were together. It was formed by the etheric of the Sun on the one-third of the Earth marked in red (Figure 55, Chapter 7). The first substances precipitated there formed the continental core. A hot, liquid basalt layer continued to exist on the other two-thirds. The double lines indicate that high mountains were folded there. The single lines show fracture zones without high mountains. It was not until the end of the Mesozoic era that the ocean floor solidified and crystallized. The precipitation of ocean water began at the end of the Mesozoic era and occurred mainly during the ice ages. This has been proven many times over by the worldwide exploration of the ocean floors! The development of mammals from marsupials onwards also provides further evidence.

How thick rock layers formed on the Earth's surface can still be observed today in the life processes of the lowest-living bacterial-plant organisms. In confined depressions with warm, stagnant water, algae mats form the so-called plankton in an oxygen-free environment without air. Algae

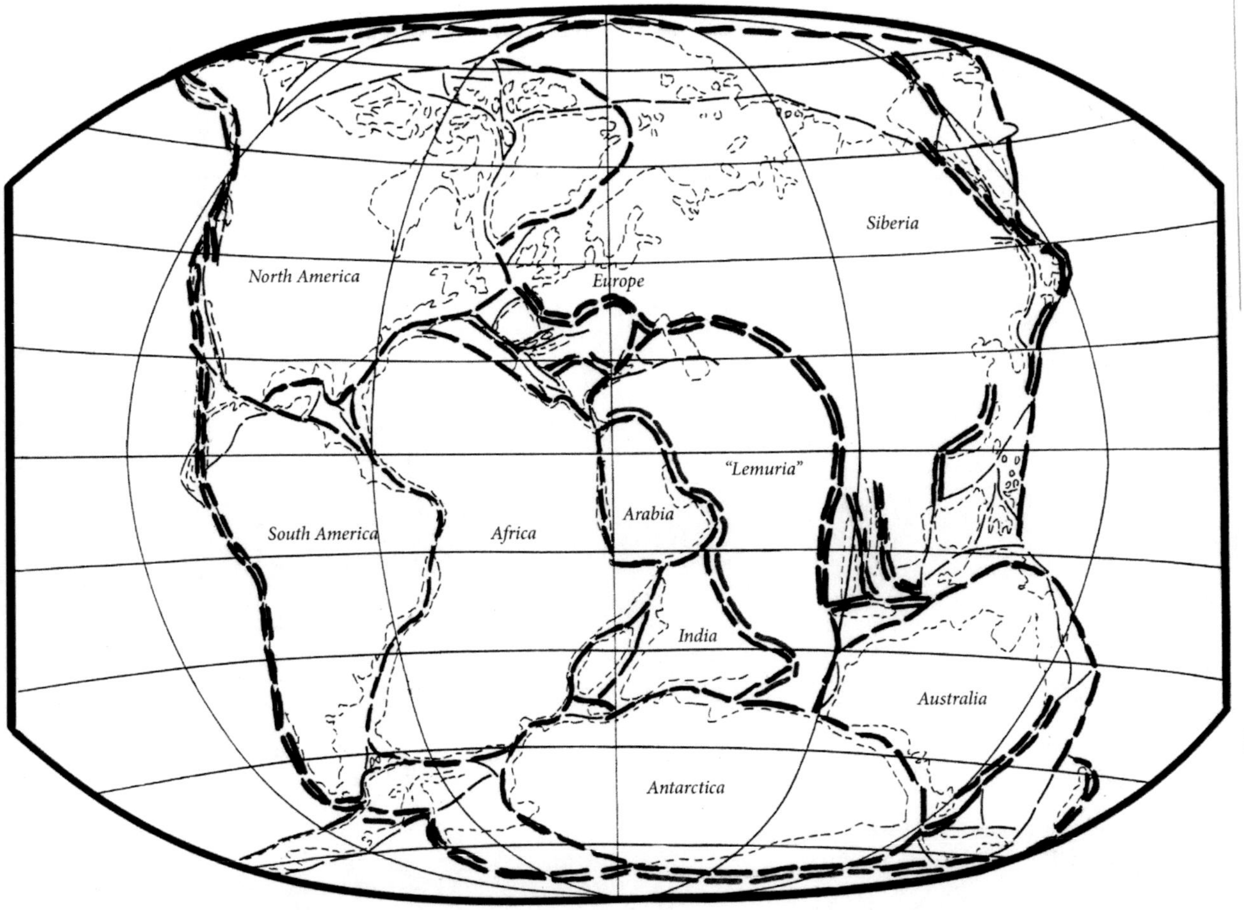

FIGURE 80 The probable structure of the large granitic continent (Pangaea) at the end of the Paleozoic era (after Owen 1983) would have covered about one-third of the spherical surface of today's Earth. The view is from the Sun to the granite-gneiss crust of the supercontinent during the Lemurian epoch. It is remarkable that the area of the current Palestine very precisely sits in the middle. The axis around which rotation of the Earth began in this epoch would have been orientated roughly as it is today.

are lower plant organisms formed by bacteria. Figure 81 shows such algae mats on a shallow lake with warm water in Landmannalauga, near a volcano in Iceland, and Figure 82 shows a closeup of this situation nearby. Figure 83 shows algae layers at a cold-water iron spring north of Berlin, originating from surface water containing brown iron. Their metabolism consists of the densification of carbon dioxide (CO_2) from the atmosphere, which produces the hydrocarbon methane (CH_4). In Figure 84, you can see a thick algae layer on a river that continues growing upward and constantly dying off underwater.

Bacteria reproduce constantly; a bacterium divides about every half hour so that, in purely mathematical terms, one-quarter quadrillion offspring could be produced in a single day. Nevertheless, the total quantity is kept in balance, which is an overall expression of an organism. The formative activity of the etheric body can be clearly observed here. If the water is somewhat calcareous, thick layers of limestone can form quickly. In a volume of humus-rich soil, for example, there are estimated to be 100 million algae, and in sandy soil, 100,000. Algae can withstand temperatures between +140 °C and -250 °C. Such bacterial algal mats formed on a large scale during the Mesozoic and Cenozoic eras in the still hot or warm water of the dying life sphere.

Bryozoa, another group of the lowest, living bacterial-plant organisms, are similar to algae. They are described in Part I, Chapter 4, The Calcareous-Clayey Sedimentary Rocks, and shown in Figure 31, as they occur in the fine-grained chalky limestones, the rhythmic layers of flints. Like corals, they form pouch-like bodies with male and female egg cells that constantly reproduce, forming new fertilized germs carried away by the water and forming new bodies. As a result, thousands of new animals are formed quickly, similar to algae. In Figure 31, fine bodies have disintegrated due to the rising water current. However, they do not concentrate lime but siliceous matter with a water content of about 10%.

FIGURE 81 Algae mats at a warm water spring in Landmannalauga, Iceland.

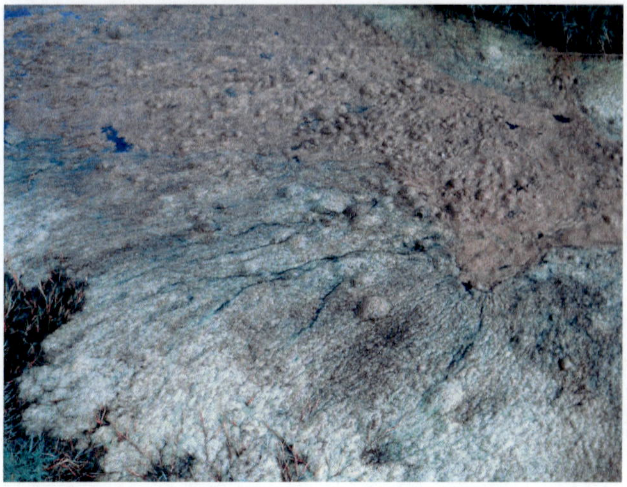

FIGURE 82 Closeup of algae mats at a warm water spring in Landmannalauga, Iceland.

FIGURE 83 Algae at a cold-water iron spring north of Berlin, Germany.

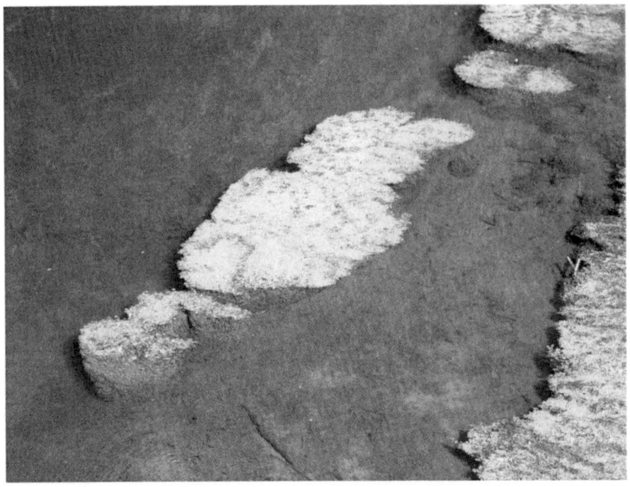

FIGURE 84 A layer of algae on a river showing present growth and dead old growth that has accumulated underneath.

Similar processes having a rock-forming effect worked in the Earth's circumference during its history. An important difference between basalt and granite is that the dark mica biotite is missing: (K(Mg, Fe)3 [OH2 I (AL, Fe) Si3O10]). The feldspars are sodium, potassium, and calcium silicates; the mica also contains potassium and aluminum silicates. The average quartz content (SiO_2) is lower in basalt at around 45-52% than in granite at 65-75%. But these minerals are not composed of chemical elements! The elements only appear through laboratory analysis! Even at high temperatures, etc., minerals can never be formed from chemical elements! The minerals and the rock sequence clearly show that the Earth with the granitic continents could not have originated from the basalts.

Above all, the hydrogen and oxygen content is missing, i.e., the water (H_2O)! Where do the unimaginable 1.4 billion cubic kilometers of salt water that fills today's ocean basins come from? All these minerals have an oxygen component, i.e., air. The terrestrial atmosphere, which can reach the stratosphere at 20-50 kilometers, contains about 21% oxygen and 78% carbon dioxide, while the planets have 95% carbon dioxide and 0.03% oxygen. And even more so, the living nature kingdoms with human beings only exist on Earth. This all points to the fact that from the beginning, an airy, warm sphere permeated with life was at work, which I have called a *life sphere.*

Rudolf Steiner gives many different descriptions of the life sphere; for example, about the volatile liquid silica substance, he says:

> *A person could see that which was in a certain sense already within this albuminous substance —finely distributed but everywhere with a tendency to crystallization — how that which was present in a finely distributed condition as silicic acid was in reality a kind of sense-organ for the Earth and could take up into itself from all sides the Imaginations and influences from the surrounding cosmos. Thus, in the silicic acid contained in the earthly albuminous atmosphere, imaginations were concretely, externally present everywhere.* (CW 233, December 26, 1923)

And in another passage Rudolf Steiner says:

> *Thus, we are led back to the Earth as an organism through spiritual science – not to the original state of the Earth as a dead mass. It was not the lifeless that preceded the living, but the other way around, the living preceded the lifeless. The lifeless mineral is a product of secretion, just as our bones are a secretion of our organism. In the same way, all rock is a product of the secretion of our Earth's organism, and its spiritual and soul processes have caused it to produce such secretions. Granite is the oldest secretion; it was originally contained in the great living being of the Earth. To develop, it needs to be separated and fall to the Earth. In the course of the Earth's history, the destructive processes became more and more prevalent until today when only chemical and physical processes are still at work. The life we find on Earth today has been sent ahead of the material masses to develop on them.* (CW 60, February 9, 1911)

CHAPTER 11

The Ordering of Time *in* Earth's History

The relationship between time and space is a characteristic feature of life. Rudolf Steiner introduces this topic with the words:

> *When we look at the ego in its life between birth and death, we cannot remain in space. We come out of space and into the pure flow of time. Time must always be experienced through processes in space. If you want to experience time, you look at the clock, for example, as an earthly image of the course of the Sun; you have an idea of time because the positions of the hands or the Sun changes, i.e., you have an idea of time through the fact that spatial things are changing in front of you. But you really only experience time in the soul, where time is a reality."* (CW 236, JUNE 4, 1924)

Just as we need years to order historical events, we need a measure of time for rocks that can be valid for Earth's history. At first, the order in years fails here. Every number from the geologic time scale with its millions of years remains unimaginable and abstract, which is not based on the Sun's orbit, as the years are counted. The time factor appears as a limit to knowledge. As a relative measure of time, however, we initially find images of diverse rhythms of different colored layers and that the rock sequences usually become ever younger from bottom to top. We experience rhythms in all life processes, in the heartbeat and breathing, in the day and night of the Sun's course, which reveal a trend. The Danish scientist Nicolaus Steno formulated his so-called *basic stratigraphic law* in 1669, now termed the law of superposition. For example, the following concepts can be stated in general terms: the granite layer above the basalt is around 4,000 meters thick; crystalline schists are thousands of meters thick; sediments are hundreds of meters thick; or unconsolidated rock is tens of meters thick. This suggests that as the Earth aged, the separation of rock substances from the life sphere was first rapid and then increasingly slower. To answer the question of the age of a stone, it would, therefore, be necessary to classify it in a coherent series of events that ran parallel to its development. Interdependent sequences of events make it possible to compare historical and geological processes in terms of their relative temporal rates.

Another time scale can be seen in the annual rings of trees, for example, in Figures 85 and 86. If they are measured, counted, and compared using their thickness, *dendrochronology* produces sequences of wider growth in spring and narrower growth in fall. In this way, it has been possible to date old timbers from ancient oak trees or Egyptian tombs, dating them almost seamlessly to around 5,000 BC. Another plausible consideration was first offered by Brandt. He tested the time relationships against the facts of human prehistory and its consequences.

> *As a plausible solution to the puzzle of the almost complete lack of population growth, the cultural-technical stagnation, the small number of stone tools left behind, the relatively few settlement sites, or the example of a few cave excavations demand a drastic shortening of the time frame of human history compared to conventional ideas. Instead of the radiometrically dated two million years, history probably only lasted a few thousand years.*

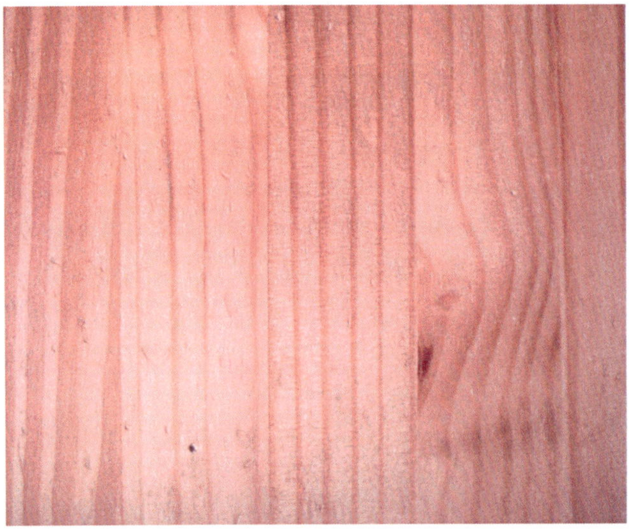

FIGURE 85 The time scale in annual tree rings.

If more human beings had lived, there would be many more remains worldwide to prove it.

Another measure of time is provided by inland ice drilling on Greenland and in the Antarctic. The ratio of oxygen isotopes was correlated with this, which counted over 40,000 annual rhythms of ice layers. This would completely contradict today's view, as the beginning of the ice age rhythms is calculated to be around one million years ago according to radiometric dating.

The oldest plants in geological history are the conifers, such as the common pine, *Pinus silvestris.* Figure 87 shows the ends of its branches, which exhibit three stages of female cone development: the new shoots that carry the female strobili in the present year can be seen developing upward near the two-year-old green cones on the left, and the two wood-colored cones near the bottom-center that are the oldest, three-year-old cones. The female strobili (not clearly shown) form on the new shoots of the present year and become the second-year green cones. The numerous small yellowish pollen bodies on some of the branches are male. From them, the pollen is carried by the wind in the spring of the

FIGURE 86 The time scale in annual tree rings.

FIGURE 87 The common pine, Pinus silvestris.

present year to the female strobili. In conifers, the seeds lie uncovered and compressed between the scales of the cones and are grown together with triangular wings used by the wind to carry them away. The cones are uncovered, i.e., without pistil, stigma, and perianth, as with the angiospermous flowering plants. The three stages of cone development were imprinted in the conifers at the end of the Paleozoic era when the *Pinus sylvestris* trees were first formed. The triennial rhythm was imprinted in the species because the Earth's orbit around the Sun at that time took three of our present years to complete rather than the one we now experience.

The development of the trees continues in the Mesozoic era with the biennial ginkgo trees, *Gingko biloba*. Instead of needles, the earliest leaf forms from the Carboniferous onward already have transitioned to fewer divisions (Figure 88), and up to the present-day representatives, mostly with one division, up to closed leaves, as in the annual deciduous trees, which only developed at the end of the Mesozoic era. The ginkgo trees are dioecious and have green flowers at the ends of the branches in summer, and small male and female cones. The numerous stamens contain pollen and are spread by the wind, as with conifers. Female cones form at the ends of branches, from which orange-brown stone-like fruits form in the autumn. In the transition to herbaceous flowering plants, the *dandelion* (*Taraxacum officinalis)*, like the ginkgo trees, is biennial (Figure 89) and developed during the Mesozoic era. In the first year, the plant only produces green rosettes of leaves, which die back in winter. In the second year, the yellow flowers are formed, which develop into *puffballs*, where the seeds hang from the fine threads and are dispersed by the wind, as with conifers. At the end of the Mesozoic era, deciduous trees emerged in the evolutionary series of plants. The example of oak leaves *Quercus robur* (Figure 90) shows the effect of the Earth's forces through the plant's etheric body. These forces go through the stem into the *leaf veins*

FIGURE 90 Oak leaves of the Quercus robur are deciduous trees.

FIGURE 88 Leaves from the dioecious Gingko biloba tree.

FIGURE 89 The dandelion (Taraxacum officinalis).

FIGURE 91 Diagram of evolutionary directions using the example of the animal kingdom. From the continually ascending macroevolution of the type, there separate off groups, from which species separate off stepwise and related to one another. They then alter continuously in the succession of generations, termed devolution or microevolution. Their forms harden to the point of creating fossil remains. Given certain conditions, they can further subdivide into races; in the end, their lines end in extinction.

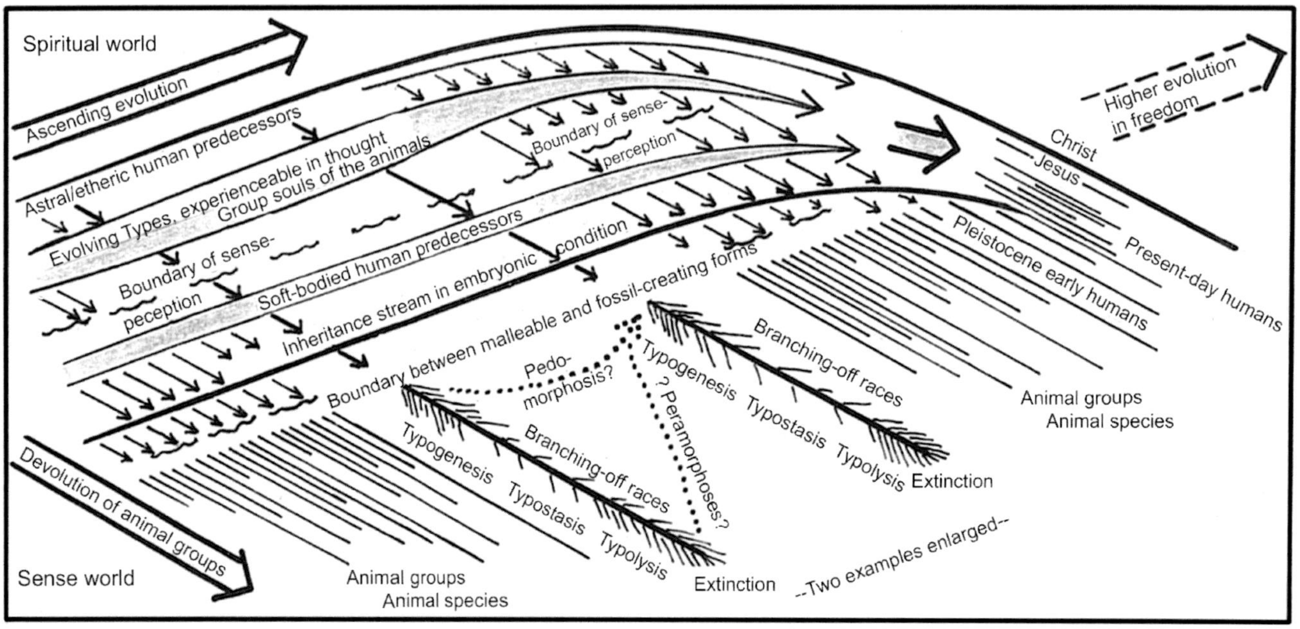

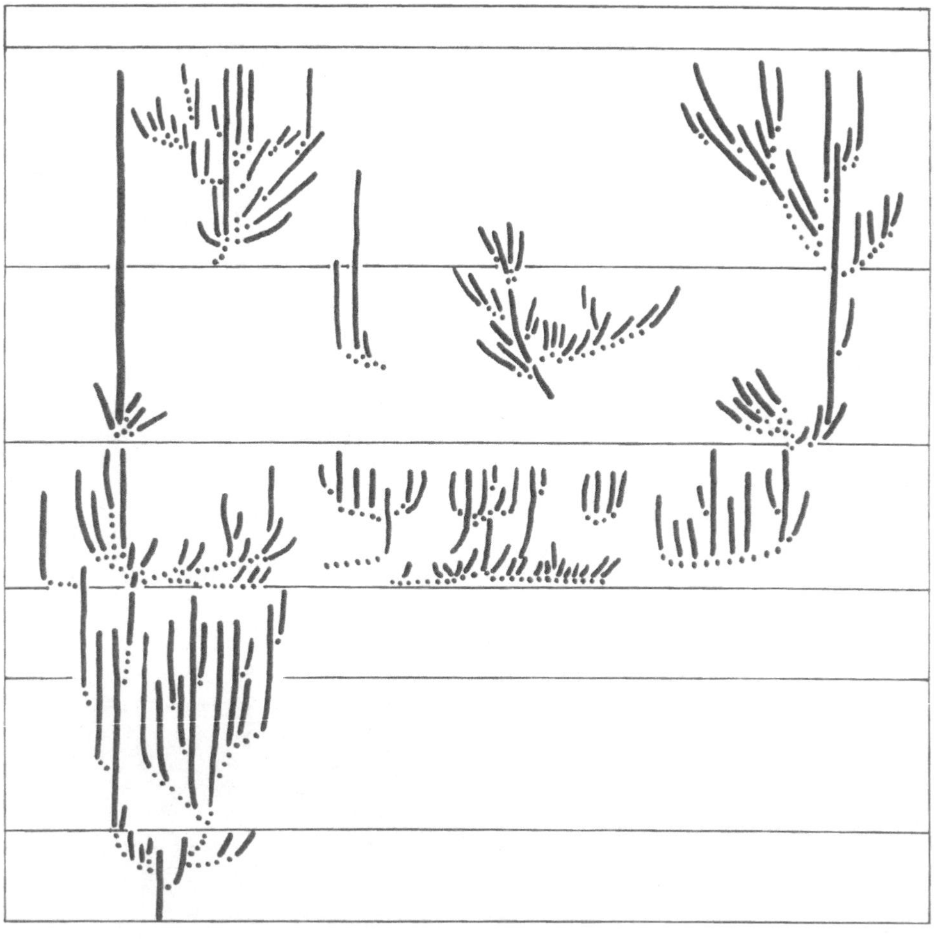

FIGURE 92 The appearance in time and relational connection (points) of 186 families (black lines) of ammonites in roughly 1500 genera and their extensiveness in time. Primarily from Müller 1961)

to the tips, while in the opposite direction the astral cosmic forces of the Sun cause indentations.

The fossilized remains of animals are another time scale for Earth's evolution. In the 19th century, fossils were used to divide the ages of Earth's history; the Paleozoic era begins with the oldest finds, and the Mesozoic and Cenozoic eras are marked by mass extinctions of most plants and animals. Particularly important today are the animal microfossils, unicellular living organisms that are already individualized and occur in enormous quantities. They can now be used to establish the age relationship of sedimentary rock layers over long distances worldwide, often to the nearest meter. This is mainly used in routine operations by the oil industry, where each meter of drilling often costs large sums of money. Overall, a time signature is thus imprinted in the rocks of the Earth's crust in several ways, which, however, contradicts the radiometrically measured data. Thus, the geological development of animals can lead from the lowest species to the highest groups, the vertebrates. The anatomical structure of human beings follows directly from this.

The diagram in Figure 91 shows this principle, discovered around 1890 by the French paleontologist Dollo and has since been referred to as *Dollo's rule*. However, the causes have remained unclear to science despite many attempts, microscopic examinations, and hypotheses to *prove* a continuous developmental series contrary to fossil evidence. It is not accepted that animal species can become completely extinct because how else could completely new species arise? One suspects a further development, such as juvenilization or *paedomorphosis*, or continuous survival, a *peramorphosis*, instead of a

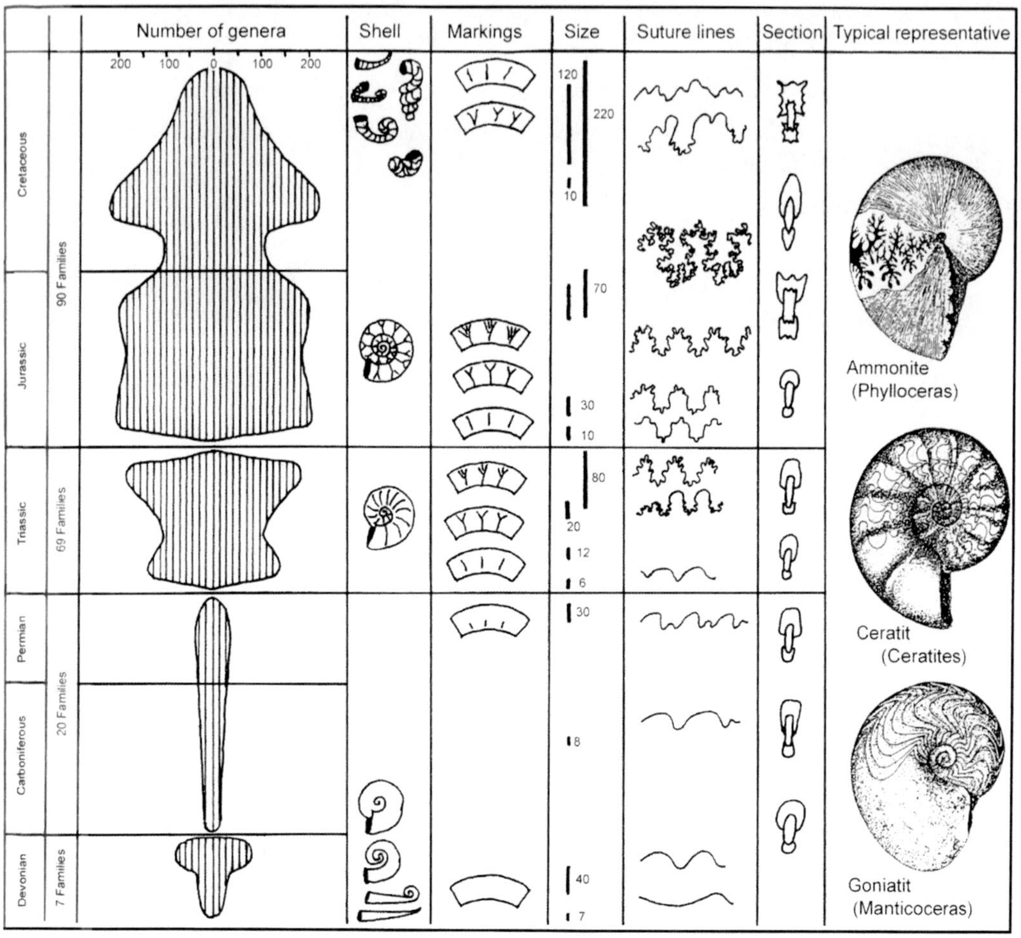

FIGURE 93 The evolution of various distinguishing characteristics in time shows an overall evolution in three developmental phases that build on one another; the far right is a typical representative. (Figures on the right from Schad 1989).

definitive extinction. But an animal species or genus dies out completely, while the human being is led into the next incarnation through reincarnation by the higher spiritual beings. These phenomena of *devolution* are precisely described, which lie below the *boundary of sense perception*, which is drawn as a wavy line. The higher development or evolution has been added by Rudolf Steiner and can continue through *Christ* for the human being in freedom, which is compiled in this comprehensive picture.

Remarkably, Dollo's rule is proven particularly clearly by the development of the ammonites! In Figure 92, the lines show the chronological occurrence of approximately 1,500 studied genera over four long epochs. The dots indicate an ideal correlation, which cannot be confirmed by the findings shown in Figure 93. This temporal development from the Devonian period to extinction is again represented by the number of genera. It is not continuous, as the proponents of the hypotheses assume, but after each extinction, a new stage of higher development begins! This is confirmed by the shape of the shell, the sculpture, the size in centimeters, and the longitudinal section, each in a different way. On the far right are three typical representatives of similar size from these respective epochs, with the fine *lobe line* in particular. This is the imprint of the chamber septa, which shows the individual life from the first to the last chamber. It reappears in each epoch of life as the animal grows larger and outward. In the end, decadent forms appear, which in some respects are similar to the forms at the beginning, in the Devonian, but appear irregular, as if out of form.

Using the example of the ammonites, all these characteristics clearly demonstrate that continuous further development, a *paedomorphosis*, is not possible because each life cycle begins anew from the first chamber! Such an embryonic form cannot usually be fossilized in other animals, which is what the hypotheses are based on! Thus, Dollo's rule for groups and species of animals is fully confirmed by the ammonites, and all such hypotheses are invalid! The overall forces of the group ego are invisibly at work in animals, guiding the continuity of overall development. In the case of human beings, one can learn from anthroposophy how their ego can continue to develop through incarnations, which does not happen in the case of animals.

CHAPTER 12

Radiometric Time Measurement *in* Millions of Years

The upper part of Figure 94 shows the annual growth rings of a cockle, *Cardium sp.* This also confirms Dollo's rule because it is inconceivable that the next generation can develop from the last stage. The growth curve of the life process and the curve of soul-spiritual ripening in relation to the half-life decay curve of radioactive isotopes can be seen in the lower part of Figure 94. These curves show their polarities and similarities in that they are mirror images of one another. The initial quantity of an isotope decreases exponentially over time. In other words, looking at it from the opposite direction, the amount of substance that has not yet decayed becomes exponentially larger. These half-lives measured today are taken as constant on the assumption that the physical processes and the Earth's orbit around the Sun have remained constant since the crystallization of the oldest Archaean gneisses parallel to the horizontal x-axis. However, this does not apply to the other possibility that the Earth was a living being. I would like to thank the Belgian physician Luc Vandecasteele, the Swiss mathematician Heiner Frei, and the German mathematics teacher Dieter Kötter for their thorough considerations and corrections on the questions of dating.

The most important method for determining an *absolute* time scale for the Earth's history is determining the decay series of isotopes of radioactive elements. The number of years can be calculated by determining that half of a radioactive isotope decays into a stable form at a consistent interval under constant conditions. The ratio of the initial and decay products is determined. Some half-lives are very long – up to 4 billion years, while others are much shorter. For example, uranium-238, which is important for determining geological time, decays to thorium-234 in 4.4 billion years, which then decays to protactinium-234 in 24.1 days, and this decays to uranium-234 in 6.7 hours, and so on, until the final product is produced, lead-206. The high numerical values of the half-lives in the rocks indicate that the Earth is very old. However, the larger such numbers become, the less content there is for them regarding actual events! They become more and more a mere calculation variable.

The growth curve of life processes shows the physical-etheric development of organic life. In the beginning, considerably more developmental steps, substance turnover, and growth occur in the same unit of time than later, in old age, when all processes become slower and slower. The physical developmental steps of human beings are looked at embryologically in days and weeks, infants in months, and children and adults in years. In animals, exemplified here by the cockle, the life process is similar: first faster, later slower, and again later slower still. The growth strips of the cockle show how less and less substance is

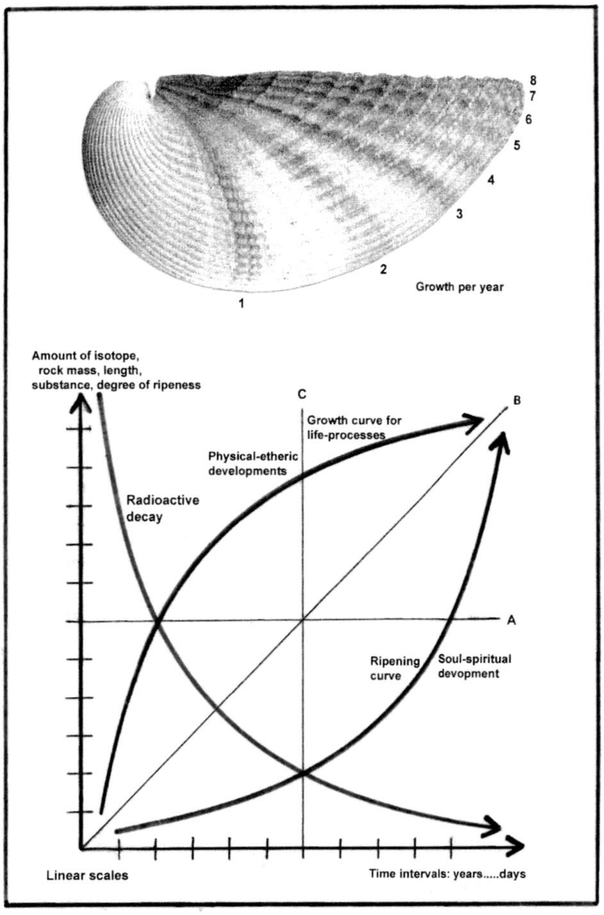

FIGURE 94 The annual growth rings of a cockle, Cardium sp. Above, and growth and maturation curves below. The growth curve of the life process and soul-spiritual ripening in relation to the half-life decay curve of radioactive isotopes is in the lower half. The curves show their polarities and their similarities in that they are mirror images of one another: the growth curve and the decay curve reflected about line A parallel to the X axis, the soul-spiritual ripening curve and the physical-etheric growth curve about line B, and the soul-spiritual ripening curve and the radioactive decay curve about line C.

accumulated each year. Therefore, it can be assumed that in a living Earth, the separation of the rock substances from the life sphere is also rapid at first and then slower and slower. This progression can also be represented by a growth curve. Of course, the thicknesses of the rocks vary within wide limits, but the following can be generally stated: the basalt layer above the peridotite is about 10,000 meters; the granite layer, about 4,000 meters; crystalline schists, thousands of meters; sediments, hundreds of meters; unconsolidated rock, tens of meters. From this, it can be assumed that the Earth is much younger than it appears through radiometric dating: Precambrian era-six billion years; Paleozoic era-291 million years; Mesozoic era-186 million years; Cenozoic era-65 million years. The time steps towards the present prove that the Earth's mineral crust also became increasingly older and is now dead and crystallized.

The maturation curve shows the soul-spiritual developmental steps per unit of time; for example, learning proceeds slowly at first and later becomes more intense and inward in an ever shorter time – spiritually speaking. It was the same in the cultural epochs of humankind. Rudolf Steiner describes this as the occult-astral stream of time running backward, which works from the future into the present where it encounters the past" (CW 262, February 4, 1913).

Let us consider this topic further with the following question. What is the relationship between these three exponentially trending curves, which are *interdependent in content*? The growth and decay curves are similar in principle; they are mirror images of each other. As line A shows, they begin with many steps, which decrease through time. Radioactive decay, on the other hand, decreases exponentially because only half of the remaining substance decays from a large initial quantity per unit of time. Organic growth also decreases exponentially. In the growth curve, however, it is not the amount of substance added per unit of time that is plotted against time but the absolute length, weight,

RADIOMETRIC TIME MEASUREMENT IN MILLIONS OF YEARS

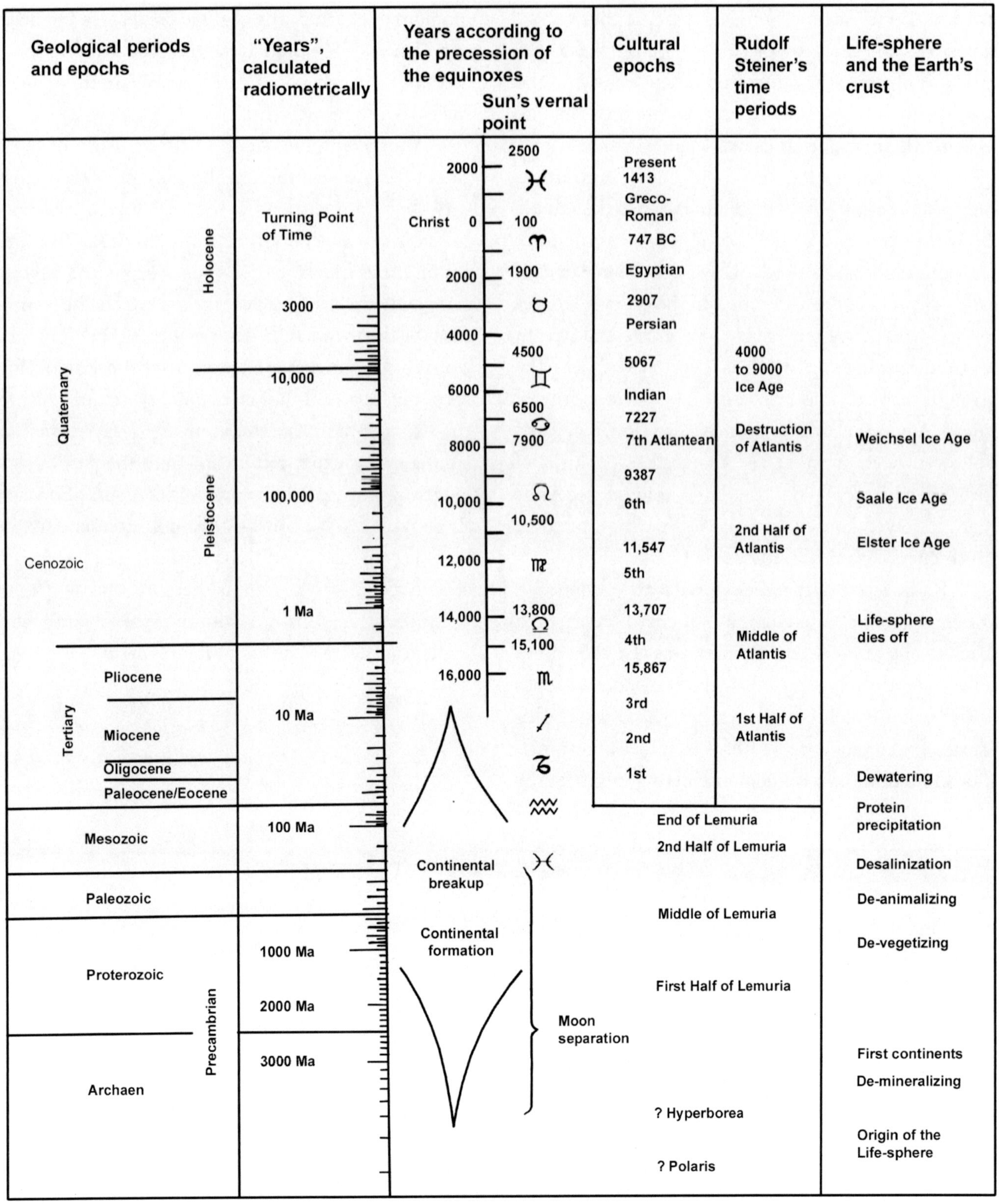

FIGURE 95 The arrangement of the Earth's development history used by natural science and spiritual science, as presented in the book "Essays from Earth's History," by the author.

thickness, etc., achieved in each case. If these characteristics for the growth curve were on the y-axis per unit of time, the curve would be identical to the decay curve. The growth curve is mirrored on line B with the maturation curve. It shows an acceleration toward the future; the growth curve, in contrast, shows a deceleration. The life processes in the etheric body thus prove to be the polar opposite of the maturation in the astral body. If we then think of this soul-spiritual curve as running in the opposite direction, it would be an image of the occult-astral stream of time running backward, as mentioned above. The growth or maturation curve is the polar opposite of the decay curve. Maturation occurs slowly at first, followed later by more and more steps per unit of time; in radioactive decay, a large amount decays at first, followed later by less and less per unit of time. Both curves are mirrored by line C.

Radioactive decay is measured by the amount of gaseous isotope. Therefore, it can only be measured back to the point where the rock was last cooled below a certain temperature so that no more isotopes can escape, which is the so-called *Curie point.* This means that a sequence of events solidifies and transitions into duration. Radiometric dating indicates the moment of death of the organic processes. The decay curve should begin at the end of the growth curve, meaning the decay curve cannot say anything about the earlier life processes.

Another problem is that the half-life of decay can only be determined in the laboratory today for a very short moment and is then calculated backward in Earth's history. But the result cannot say anything about these much longer times. From the asymptotic end of a decay curve, the rise of the beginning must be determined. Physics assumes that all natural constants are invariable throughout the Earth's history. But how did radioactive decay begin? It is not the exponential progression of the decay curve that simulates the extension of time into the past but the length of the half-lives under the assumption that these were always constant. Rudolf Steiner explains that the physical conditions in the Earth's early history differed from today's because the aggregate states today are separate from the etheric forces with which they were together in the Hyperborean epoch.

> *Physics is not eternal. It is no longer valid for completely different kinds of realities. A reality where the gaseous is internally luminous is a very different reality from that where the gas and the light are*

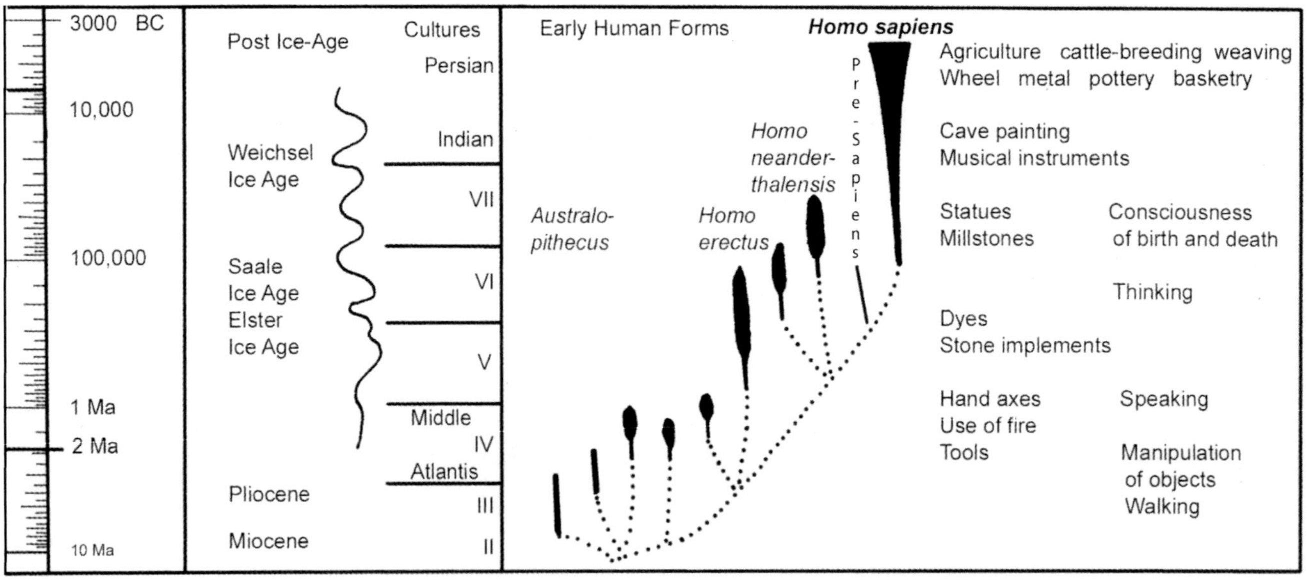

FIGURE 96 The sense-perceptible geological development of the physical human being.

relatively independent of each other. Where gravity ceases, and everything flows outwards, our physical worldview ceases" (CW 321, March 13, 1920).

An absolute age is still spoken of primarily by popular science and in museums. In scientific research, the term radiometric dating is more correctly used. This method can be successful with the different solidification times of basaltic and granitic rocks containing no fossils and assigning these values to the geological epochs through the differences in the decay series of radioactive isotopes.

Rudolf Steiner explains his concept of time in terms of the interrelationship between *eternity and development*, i.e., between the duration in the soul-spiritual realm and development in the transient nature of the world of phenomena, which is shown in Figure 95 (CW 184, September 15 + 20, 1918). He only used the concept of the year in today's sense, going back to 3,000 BC, i.e., when the historical record begins in the Babylonian-Egyptian epoch. To relate even more distant cosmic and human events to each other, he turned to the *Platonic world* year in 1910, which comprises 25,920 years in which the vernal equinox of the Sun passes once through the entire zodiac (e.g., in CW 349, February 17, 1923). Today, this vernal equinox is in Pisces. Rudolf Steiner goes back to the middle of the Lemurian epoch when the Sun was also in Pisces. This would take us back to the second half of the Lemurian epoch. It is probably only from then on that the Earth's current orbit around the Sun can be reckoned with, giving the year its current length. Further back, the year loses its temporal reality in today's sense. This corresponds to the times of plant development.

Figure 96 illustrates the sense-perceptible geological development of the physical human being, beginning with the oldest bone found at the end of the Tertiary period in the Cenozoic. The dark black lines indicate documented finds, and the rows of dots indicate their ideal connection, as already described in Figures 92 and 93 for the ammonites in the previous chapter. Even the four groups of the ancient relatives of humans (*australopithecines*) have recently been classified differently. Previously, the earliest finds were made in South Africa and the Oldoway Gorge in Kenya. In the meantime, even older finds have been made at Lake Chad when the Sahara was still a forested steppe, and the very oldest bone finds come from the areas of Palestine! Previously, the oldest finds came *out of Africa*, but that rule is over! The rows of dots show the suprasensible development of today's human beings (*Homo sapiens*), which shows the three spiritual stages of development – walking, speaking, and thinking – leading beyond the animal stage and confirmed by archaeological finds. The earliest human form, *Homo neandertalensis*, is from the Neander Valley in western Germany, with the bone finds shown in black. This find also points to the three spiritual stages of development. Since then, similar findings have also been made in other countries. The pre-Sapiens line shows an assumed intermediate form. This whole period is a hotly contested archaeological field when it comes to proving the hypotheses of the ancestry of human beings. However, how it has developed since the beginning of the Earth's evolution with the nature kingdoms will be shown in more detail in Part III.

PART III

The Separation *of the* Nature Kingdoms *from the* Human Being

CHAPTER 13

The Polarian Epoch: Old Saturn

The development of the Earth – with its four nature kingdoms, minerals, plants, animals, and human beings – begins with the state of Old Saturn, in which the physical body of the human being was conceived. Initially, this was a spiritual connection of forces in the substance of warmth. At the present stage of the Earth, the body is in a mineral-physical state and remains behind at death. To this physical body are added three spiritual aspects of being: on the Old Sun, the etheric body is added as the life or formative body, which builds up the physical body and today remains connected to it during sleep. On the Old Moon, the astral or soul body followed, and during the Atlantean epoch on the earthly plane, the ego or "I" followed with the consciousness soul. This results in seven stages of being. On Earth, all four nature kingdoms have, like the minerals, a physical body; the plants have, in addition, an etheric body; animals have an etheric body and an astral body; the human has an etheric body, an astral body, and an ego or "I," which can develop further through repeated incarnations.

The Polarian epoch, a repetition of Old Saturn, is described by Rudolf Steiner as a state of warmth.

> *The beings of the First Hierarchy, Seraphim, Cherubim, and Thrones, work together as a choir, weaving into one another, doing, thinking, and feeling in one another, radiating warmth into the universe. [...] But all of this is soul and spirit, but in such a way that it is a physical experience in our senses, and this warmth formed the Saturnian existence.* (CW 233a, April 1, 1924)

Figure 97 (see next page) is a picture by Rudolf Steiner that shows this state of soul warmth and is accompanied by the following comment:

> *Around our Earth, we find within all that belongs to our Earth something like live members of a great living being. On the Earth's surface, however, the temperature is so great that life cannot develop, but life is developing around it, life that wants to descend but cannot. Down there, the fire process of high heating first absorbs what the living Earth secretes out of itself, just as our living organism secretes the solid components, the bones, out of the soft parts. Now we look at the formation of granite and say the material that granite contains, mica, feldspar, and quartz, was originally dissolved in the great living being of the Earth. For its development, it needs the fact that it can get rid of these*

FIGURE 97 The Weaving of the Light by Rudolf Steiner, 1911. "Then, in the gentle ether red of spirit worlds, I try to densify what is invisible." Words by Thomasius in The Souls Probation.

FIGURE 98 The evolution and devolution of the human being, from a lecture sketch by Rudolf Steiner. "You behold in what nature-knowledge shows you that which you need to exclude from knowledge of the human being." (CW 223, April 8, 1923)

substances, it secretes them and lets them fall to the Earth. What is below takes up this separated material, forming a basic mass, a skeleton in the living organism, the Earth. And as the Earth approached the time when the granite and similar materials were separated, it was seized by nothing but destructive processes, which worked in such a way that a spiritual-soul element worked on a material one. In that organism, into which not only the substances that today are found in the plant, animal, and human organism had been worked, but also the substances that today make up our Earth's massif, everything that was present from such destruction processes caused by spiritual-soul processes poured out. These destructive processes initiated in the great living organism of the Earth that which was expelled through a process of separation, that which we have today in the form of chemical substances in the structure of our Earth, that which we do not find in organic bodies. (CW 60, February 9, 1911)

Rudolf Steiner writes about Figure 98:

> *You behold in what nature-knowledge shows you that which you need to exclude from knowledge of the human being. [...] If this in red is the original human disposition, they had to develop from that condition to their present state. First they had to separate from themselves the lower animals, then later everything that had an animal form (yellow). In this way, humans reached what they were originally destined to become. This principle also applies to elementary states. Heat is divided into inner and outer, the next state into light and airy, sound and watery, living and solid. In what nature shows*

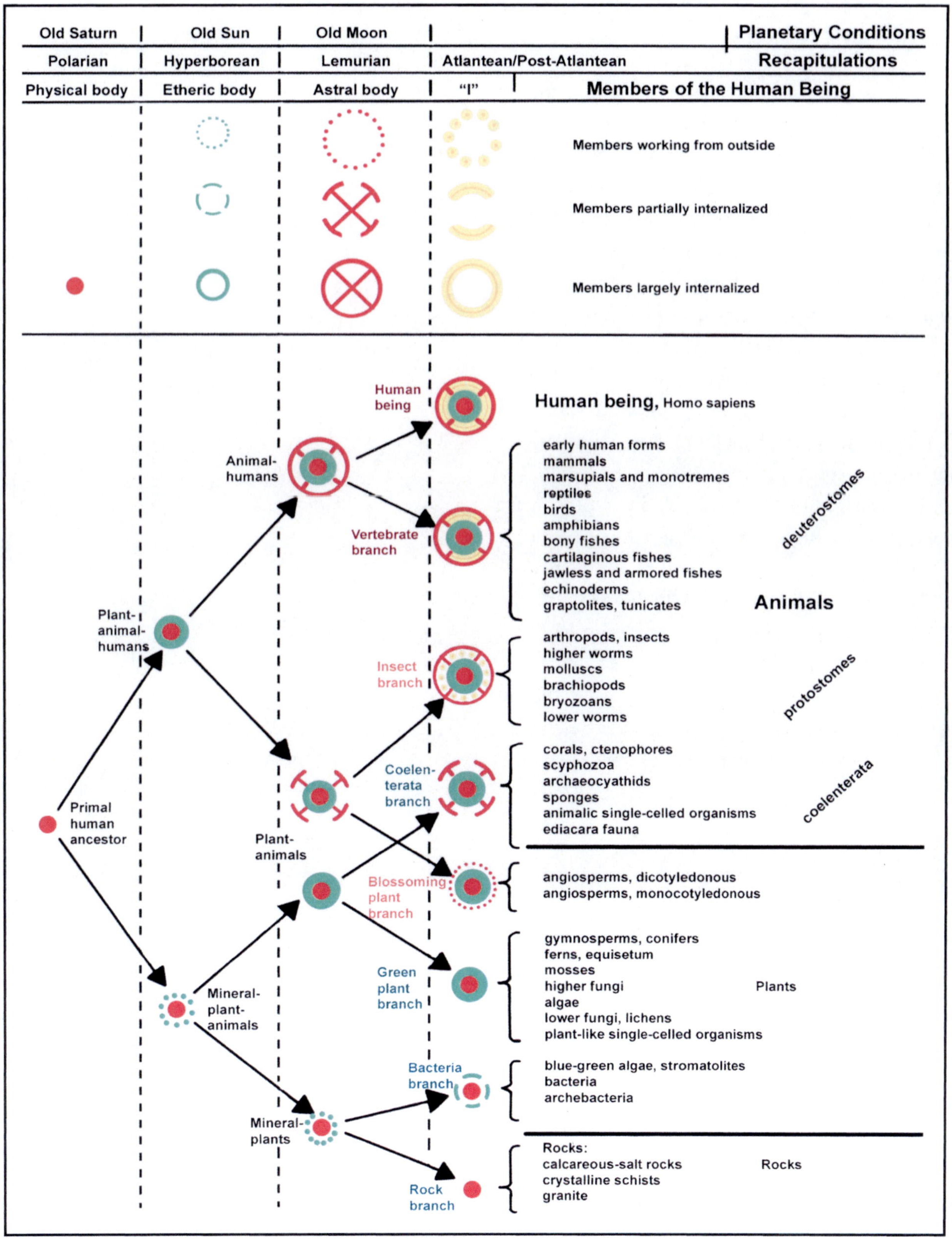

FIGURE 99 Members of the kingdoms of nature through ascending evolution and remaining behind during the planetary evolutionary stages and their recapitulations during the Earth-evolution. The internalization of the life and soul principles leads to three main groups in animal and plant kingdoms, which lie between the human kingdom and the rocks.

> *you, you see what is not spirit.* (CW 223, April 8, 1923)

In Figure 99 (see previous page), a diagram of the descent of the nature kingdoms, it becomes clear how the three nature kingdoms have separated from the human "I" in the course of the Earth's evolution. This is brought about by spiritual beings of higher hierarchies during the preliminary planetary stages of the Earth's evolution. Thus, the evolutionary developments during the descent lead to a picture of the division of the nature kingdoms from the evolving human being and shows their continuous higher development from the beginning until today. In each epoch, a denser part of the human being separated from the life sphere of the Earth and sank to the emerging surface of the Earth, while another part developed itself higher. Following this principle of differentiation, we find two nature kingdoms on the Old Sun and three on the Old Moon, with the middle one being two-fold, just as the animal and plant kingdoms on the earthly plane are each threefold. During the planetary epochs of the Earth, the higher hierarchies formed the constituent elements of the human being and internalized them.

CHAPTER 14

The Hyperborean Epoch: Old Sun

Rudolf Steiner painted a pastel sketch in 1914 for the large cupola of the first Goetheanum in Dornach, Switzerland (Figure 100). Rudolf Steiner explained the figure with these words:

> *The Elohim, the Spirits of Form, work into the Earth; the Beings of Light ray into it.*

On the left side of the figure, four red-yellow light beings ray down from above and are received by four blue beings who receive with their left hands the light beings and pass them downward with their right hands, where four brownish plant-like forms (one with a red spot) rise up from the Earth. On the right side of the figure, four rhythmically working red beings are drawn, which open downward toward delicately drawn human head-and-breast beings, from which blue rays are directed toward the Earth's surface, near what are perhaps animal beings drawn

FIGURE 100 The Elohim, the Spirits of Form, work into the Earth; the Beings of Light ray into it. A pastel sketch for the large dome of the First Goetheanum by Rudolf Steiner, 1914. (Raske, 1983)

FIGURE 101 Rudolf Steiner indicates how the quartz (blueish layer) was formed over a black interior of the Earth as the initial, still-fluid layer of rock. Section of a blackboard drawing for a lecture given by Rudolf Steiner. (CW 232, December 1, 1923)

with black lines. A fifth human being is indicated very lightly, approximately in the top center, next to the rightmost red-yellow ray of light, without a red being. This is the condition when the Earth's life sphere was created during the Hyperborean epoch. Rudolf Steiner says:

> *The spiritual, the soul of the human being, does not need life. Life only begins with the etheric body, which is outwardly shell-like. And so, life only enters the realm of evolution that belongs to our Earth after the Moon and Earth existed together. This is the colorful world of Hyperborea. And there, in the colorful splendor, forces became active and formed contours. Life inwardly rounded off the colors, angled and edged them, and called forth the solid crystalline, bringing us into earthly existence.* (CW 223, August 4, 1923)

Rudolf Steiner describes the overall situation during the Hyperborean epoch in detail, how the first two kingdoms of nature, the plant-animal-human and the animal-plant-mineral, began to form. He says about Figure 101:

> *At that time, there was around the Earth where today we have an atmosphere, something having an extremely fine fluidic nature with a consistency halfway between our present water and air. It was fluidic, but its fluidity was similar to albumen. The Earth was entirely surrounded by an albuminous atmosphere at that time (a dark kernel surrounded by a blue circle). The whole of this substance surrounding the Earth at that time was permeated from the universe with cosmic ether, which gave it life (four yellow arrows). Then the whole was pervaded by reflections of light, gleams of light, rays of light, sparkles of light. And finally, the whole was warmed through by the cosmic ether.*
>
> *The first thing that formed from the cosmos was quartz, with its beautiful form and relative transparency (above left, yellow crystal outline). This quartz penetrated the whole atmosphere, gradually depositing itself as the first primeval mountains, as the continental granite-gneiss-schist layers. And connected with this were mighty, algae-like plant formations, greening and disappearing again; deposits of the human being still completely spiritual in this aqueous proteinaceous atmosphere. And what approached was transparent, in a wax-like condition somewhat like a plant image (blue within the yellow quartz outline). Later, during the Lemurian epoch, when the Earth became denser, the mighty floating plant formations reminiscent of our modern algae emerged, which were not rooted in the ground and were changeable; they lit up, then disappeared again, and were there again. This, in turn, made them somewhat harder and denser; they were no longer mere images.*
>
> *Then, these cosmic forces worked on many parts of this albumen in such a way that it congealed so that one could see everywhere albumen congealed by the forces of the cosmos; this cosmically congealed albumen swam in the Earth's environment. These forms of cosmically congealed albumen were not merely imaginary masses of clouds; they were living things having definite forms. They were, in fact, animals that consisted of this congealed albumen, thickened to the density of jelly or even our present-day gristle (two elongated, pinkish-white areas in the yellow above the blue). Such jelly-like animals existed in this fluid, albuminous atmosphere. They had a shape that we find today on a smaller scale in our reptiles, in lizards and creatures of that kind; they were not*

so dense as these are, but they had gelatinous bodies and the power of movement. At one moment, they had long limbs, at another, these limbs were drawn back into the body. These animal beings, which the human being had separated out, were united with the calcareous, so the Earth became denser and denser at its core. Then, the siliceous is incorporated into the calcareous, but this retains the cosmic forces; the calcareous substance does not contain these formative forces, so it is a continuous upward and downward movement of the calcium in the proteinaceous atmosphere. From this, in Lemurian times, animals gradually emerge that form calcareous bones.

Then everything changed; the atmosphere became more oxygenated, and human beings came closer to the Earth. What the human being now separates appears like the descendants of the earlier plants, which took root in a quite soft Earth substance. And out of this emerged what were lizard-like animals, which today's geology still finds through impressions. These are today's lower mammals of the Jurassic and Cretaceous periods. (CW 232 December 1, 1923)

CHAPTER 15

The Lemurian Epoch: Old Moon

In the Hyperborean epoch two nature kingdoms emerged: above the plant-animal-human and below the animal-plant-mineral. After this development, four nature kingdoms separated during the recapitulation of the Old Moon during the Lemurian epoch. The human being then came closer to the emerging Earth. Rudolf Steiner drew a pastel sketch and a pencil sketch for the First Goetheanum, shown in Figures 102 and 103. The upper part of Figure 102 still resembles the one he painted for the Hyperborean epoch (Figure 100, in the previous chapter). In Figure 102, suprasensible, winged beings float in a reddish-yellow environment, representing the life sphere of the Earth. The reddish rays are directed downward, whereas plant-like figures and animal beings drawn in black can be seen in the brownish-green lower part. Figure 103 shows these beings more clearly. Rudolf Steiner says of this sketch:

> *Physically, a great diversity is shown: the mineral plants, the plant-animals, the animal-plants, and the animal-human beings. By leaving behind the formations at the earlier stages, forms have been embodied with the most varied qualities.* (CW 13, 1910)

FIGURE 102 Lemuria, pastel, R. Steiner. (Raske 1983)

FIGURE 103 Lemuria, pencil sketch, R. Steiner. (Raske 1983)

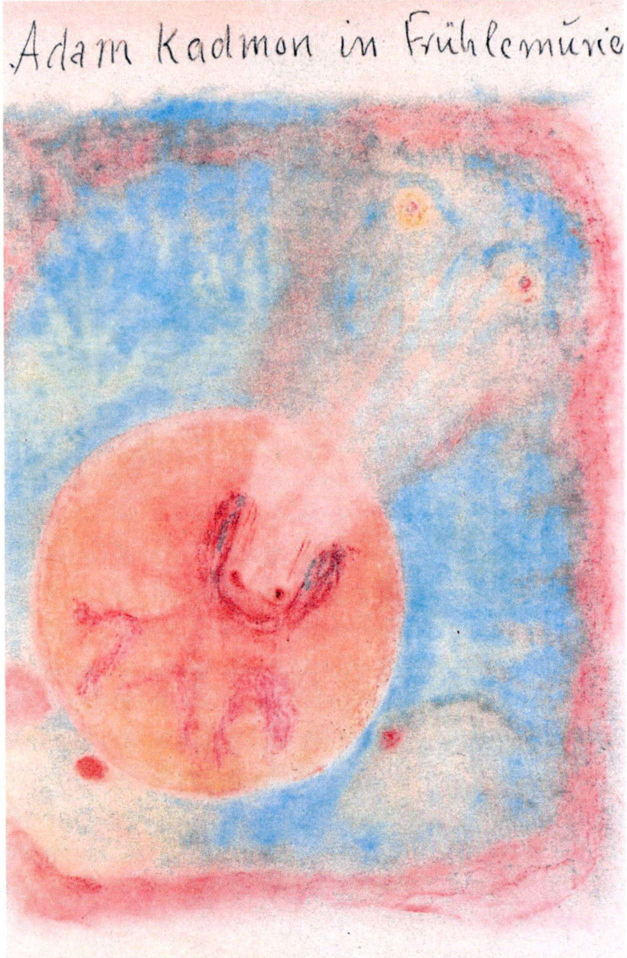

FIGURE 104 Adam Kadmon in Early Lemuria, Pastel, Rudolf Steiner. (CW K54.21 G)

These four separations from the human being result from the division of the nature kingdoms and are presented in greater detail in what follows, from bottom to top.

First, in Figure 104, Rudolf Steiner paints the picture of the human being, to which he gives an ancient oriental name. He says:

> *There is a picture of the Earth that was once there in the head of a human being when first present as a germ before birth. The Earth was once the germ of a giant human being united with the universe as the embryo is with the mother.* (CW 347, September 30, 1922, CW 54, February 1, 1905)

Spiritual beings radiate the human being onto the nascent Earth. Below can be seen two yellow, animal-like heads that were still etheric but already endowed with senses, to which he says:

> *Those nerves that go to sense organs were formed from within but not yet opened.* (CW 102, March 16, 1908)

What remained behind from the animal-plant minerals of the Hyperborean epoch became the lowest realm of Lemuria. Rudolf Steiner says:

> *The whole groundmass of the Old Moon is formed of a half-living substance that sometimes moves sluggishly, sometimes lively. One could speak of this as the realm of plant-minerals. As rock masses accumulate today, dense areas within the Moon mass formed that are comparable to hard woodlike formations or forms made of horn.* (CW 105, August 8, 1908)

From this group of animal-plant minerals, the lowest living plant organisms, the bacteria, remain and have formed rocks on a large scale.

Rudolf Steiner says of the Earth in Lemurian times:

> *On the ground, the temperature is so great that life cannot develop, but life does develop in the surrounding environment, but cannot descend. And if we look at the formation of granite we can say the minerals granite contains were originally dissolved in the great living being of the Earth. Originally, the granite material — quartz, feldspar and mica – Is in a state of dissolution within the great living being Earth. The latter needs the capability to discard these substances for its development: it segregates them and lets them fall to the Earth. What is below absorbs this separated material and forms a basic massif, a bone-structure in the living organism Earth.* (CW 60, February 9, 1911)

CHAPTER 16

The Plant-Minerals *of the* Lemurian Epoch

Figure 105 shows present-day bacterial layers on a river. The light green surface where these beings breathe is above the water; the dark green layer below constantly releases substance downward, which then unites with the sediments in the subsurface, causing the rock layers with the minerals to become higher. These processes are described in more detail in Chapters 22 and 23 concerning granite formation.

Figure 106 shows algae mats in the drainage area of a geyser in northern Bohemia. Algae tolerate great heat and cold and are thus adapted to the living conditions on Earth during Lemurian times and have continued to multiply until now. These processes are described in greater detail through pictures in Chapter 10 on the formation of the continents from the life sphere. In early Lemuria, bacteria only formed a physical body on which an etheric was active from the outside. A lifeless mineral in which only the physical is left behind only appears later as rocks on the Earth. Bacteria are considered plants because they feed partly by biological assimilation, although they are at a much lower level than plants. They fill the air, water, and soil and are subject to the etheric body of the Earth's life sphere, a higher organism.

FIGURE 105 Bacterial layers on a river.

FIGURE 106 Algae mats

CHAPTER 17

The Plant-Animals *of the* Lemurian Epoch

During the Lemurian epoch, the realm of plant animals was divided into an upper group of animals, in which the etheric body was internalized in the physical body like a plant, and a lower group of green plants, the hardy evergreens. Out of the emerging human being, the soul, the astral element could be imprinted on the etheric, through which, for example, the plant-like living mollusks came into being, which developed further as sponges, corals, ammonites, and shells, and secondly, the mosses, or the evergreen conifers, from which the coals descended into the Earth. Rudolf Steiner describes these etheric forces as

> *a being of forces that was present before anything was sense-perceptible on the Earth and came into being within it and worked on it. What is thus experienced we may just as well call the etheric or elemental basic beings or bodies of the plant and the Earth, just as we call the body with which we gain perception outside the physical body, our own etheric body.* (CW 16, Third Meditation)

Figure 107 shows a piece of brown coal with yellowish amber hardened from the resin of coniferous trees from what previously descended into the Earth. Plants cannot survive in deep water. Under shallower water without oxygen, peat is formed from which the brown and hard coals are derived. But this is where the problems begin as to how coals were formed in the first place because the thicknesses, the deposition in watery depressions, the plant remains, and many other phenomena that a thorough coal geology describes speak against it if one starts from today's observations in watery primeval forests, and not from a living Earth sphere which was still active in the Lemurian epoch until the end of the Paleozoic era. This was explained in greater detail in Chapter 3.

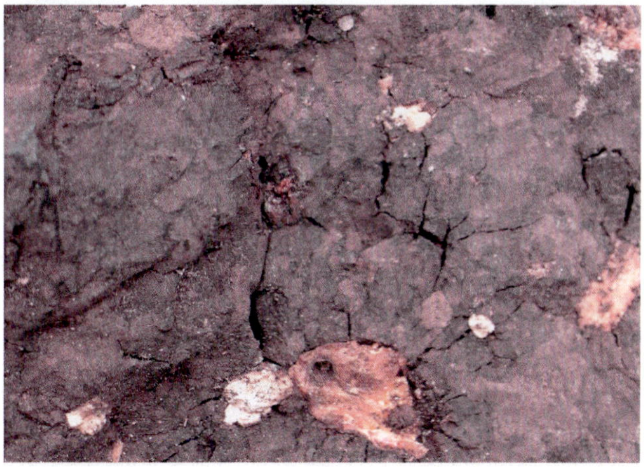

FIGURE 107 Brown coal with yellowish amber hardened from the resin of coniferous trees.

Sponges are one of the animal groups that evolved to a higher level. It is an interesting example of how the astral works in the earliest, more highly developed plant animals. I have drawn a longitudinal section through a single sponge in Figure 108. The entire animal is filled with channels. The inner cavities are lined with collar flagellate cells (*Choanozyten)* that food particles stick to, and one of these is enlarged on the right. Together, they digest the food for the whole animal. They can also swim away from the mother animal in the moving water, attach themselves, and form a new sponge. The collective movement of the flagella creates a strong flow of water in the animal, which flows in through the pores and flows out again upward via the inner cavity. No defined organs are yet formed within the soft connective tissue (dotted areas), although they contain specialized cells that perform various functions. They form canes by the budding of separated tissue parts that grow apart like plants, or they reproduce sexually through migratory cells that develop in the mother and form freely moving larvae that attach themselves and thus form new canes. This is similar to how the seeds of plants are dispersed by the wind. This animal group of sponges thus shows particularly well how the astral body, as an overarching organism, organizes the forces of the etheric body in the individual animal and leads to more sponge canes.

Rudolf Steiner says:

> *The groups of plant animals originated from a higher development in that they had internalized the etheric body at the beginning of Lemuria. They took on the astral body as they ascended further or remained at this level as a green plant. It is interesting to observe, for example, how the cosmically influenced, etheric spherical form or leaf surface dominates at first; this then changes into the spiral symmetries, as with the green plants, and is then increasingly dominated by a two-sidedness, in which the astral has an effect, and can be seen in the sponges or the corals. In nature, in the world-being in general, everything interacts with each other through the continuously active life ether and finer forces in the coexistence of animals and plants.* (CW 327, June 15, 1924)

Mollusks also belong to the group of plant animals, from which a mussel is shown in Figure 109. Ammonites also belong to this group, and further information is given in Chapter 11 on earth's history. The outer shells consist of hard, crystallized calcium carbonate, which snails can also form in the front, protecting them all around. In this way, they are similar to the Earth in this epoch, which also had a solid granite outer shell. This shows the age from the embryonic stage onwards; the outer shell of mussels, which grows a little more each spring,

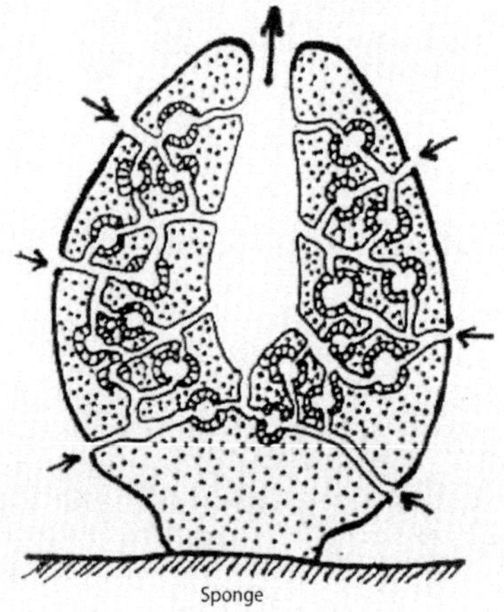

FIGURE 108 A longitudinal section through a single sponge.

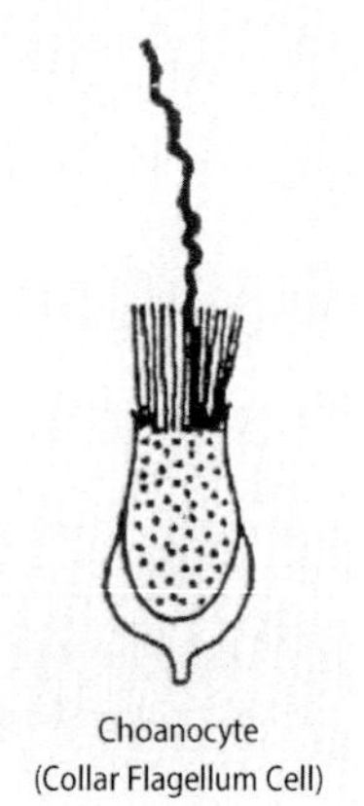

FIGURE 109 Growth per year of a mussel.

has light-colored stripes, while in the fall, it has dark stripes. In snails, the inner soft body remains united with a thread of life from the beginning, when the animal grows a little more at the front each year, forming a new sheath. Snails are a delicacy in France; in Freiberg/Saxony, there was a restaurant where, as students, we could eat fried vineyard snails – very tasty!

Rudolf Steiner says by way of introduction:

> *Around the middle of the Lemurian epoch, another developmental state of the Earth follows. It became more compact, more solid. This forced the human being to adapt to this firmer form of physical life on Earth. The human being was forced to adapt to this more solid structure of physical life on Earth. He could do this only because he handed over a part of his being to the coarser materiality. The first and most imperfect animal world originated from this part of the human being. We see these creatures, which human beings have cast off, deposited in the first layers of the Earth. They are crustaceans; shellfish humans have separated from themselves.* (CW 53, February 9, 1905)

CHAPTER 18

The Animal-Plants *of the* Lemurian Epoch

We can look at this animal-plant group with current examples like we did for the plant-animals. The flowering plants and the lower animals originated in the Upper Cretaceous period and have not died out since then, like some other animal groups. In plant life, the etheric body was internalized in the physical body out of the developing human being, and the astral led the animals, like the insects, to the flowering plants. The green leaves develop first with most flowering plants, as with the evergreen plants, but in this group, they develop further to form flowers.

FIGURE 110 Aegopodium podagraria, prepared by W. v. Goethe, (Leopoldina Edition I, 9).

Goethe describes a good example of leaf formation using fenugreek leaves (*Aegopodium podagraria)* as an example (Figure 110). The leaves are always spread out from the stems by the forces of the Earth and are indented from the outside by the forces of the Sun. Goethe describes the sequence of fenugreek leaves in his text to the picture. Only a few centimeters wide, the leaflets were probably placed on drawing paper, outlined in pencil, and then colored with ink and green watercolor. Goethe wrote this text in 1806:

> *The individual leaves are more or less indented. The single leaf separates at the bottom of the stem on both sides so that it becomes threefold; the upper of these three leaves separates again at the stem so that another threefoldness is created, and the whole leaf must now be regarded as fivefold. At the same time, you notice that the two lower leaves begin to separate at the downward-facing edge, so a sevenfold leaf appears. This separation continues until the upward-facing edge of the lower leaves is also indented and separated, resulting in a ninefold leaf that continues to be divided in this way.* (Leopoldina Edition I, 9)

The tulips (Tulipa, *Liliacaeae)* show the developmental stages from the green petals to the greenish sepals and flowers *(*Figure 111). They began to grow on the developing Earth with the forces of the etheric body and formed the most varied types of flowering plants in which the astral is formatively active on the outside through the Sun and remains on the outside.

Rudolf Steiner describes what can be observed in every plant:

> *In the seed, the plant formation is most strongly drawn together, concentrated. The first unfolding and expansion of the formative forces takes place with the leaves. What is concentrated in the seed is spatially dispersed in the leaves. In the calyx, the forces contract again at an axial point; the crown is brought about by the next expansion; stamens and pistils are formed by the next contraction; the fruit by the last, third expansion, at which point the whole force of plant life is again concealed in the most contracted state in the seed.* (CW 1, Chapter II)

When Rudolf Steiner drew Figure 112, he described this metamorphosis of the plant with the words:

> *Every spring, when we look at the plants, we see that the outer world comes from the infinite world space in the form of light and brings the cosmic into the plant. We have the root of the plant in the Earth, the stem or trunk of the plant with the leaves above that, and the flower at the top. When we look at the root, the plant in the soil is surrounded by salts, which are heavy. But it overcomes the heaviness. As a result, it becomes accessible to the light that works from above downward, opposing the heaviness; it is exposed increasingly to the Sun with its light. Fertilization arises here in it; the ovary with the germ forms so that the plant grows from heaviness to light and thus to fertilization.* (CW 350, July 20, 1923)

Flowers attract diverse forms and species of insects with their colors and scents; for example, Figure 113 shows a bee on a flower. Both groups of plants and animals developed almost concurrently during the Upper Cretaceous period. Since then, they have remained connected through the fertilization process. Interestingly, the fruits, which emerged from the flowers as spherical seeds, are spread by birds, which also emerged at the end of the Cretaceous period! One might ask: which birds

FIGURE 111 Tulipa of the lily family.

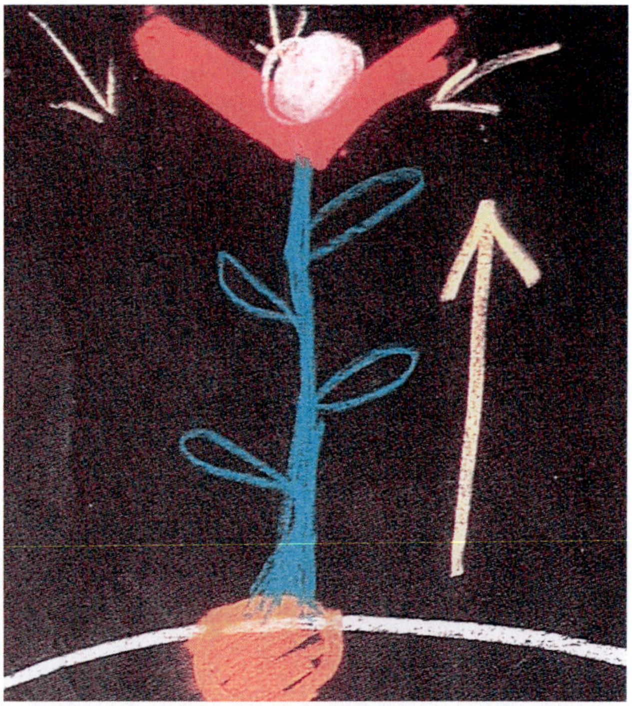

FIGURE 112 Sketch by Rudolf Steiner (CW 350, July 20, 1923)

belong to which fruits? For example, we have observed how crows flew to a walnut tree to crack nuts that have fallen to the ground. This will lead us to the next chapter about the animal-human.

As a memory of Lemurian metabolic processes, consider a well-digested humus soil or compost pile formed of a plant-animal-bacterial totality. Rudolf Steiner says that

> *life continues from the roots of plants into the Earth's soil and that there are no sharp boundaries between the life of the plant and its surroundings. Consequently, the Earth has a living soil as its own etheric, like a present-day image of the earlier life sphere.* (CW 327, June 10 & 15, 1924)

Rudolf Steiner summarizes these thoughts:

> *The spirits above human beings now work on the etheric body. Under their influence, the forces of nutritional substances were taken up by human beings from animal plants and increasingly became an inner-life activity. These animal plants can be thought of as floating, growing easily in an element surrounding them, just as the present lower animals live in water. But this element is a kind of dense vapor, a medium of water and air that moves back and forth with the most diverse currents. The animal plants appear only as densified regular forms of this element, often physically only slightly different from their surroundings. The breathing process exists alongside the nutritional process. It is similar to drawing in and flowing out the warmth as if organs were opening and closing again through these processes, a warming current through which the substances would enter and leave.* (CW 13, 1910)

FIGURE 113 A bee on a flower.

Concerning human beings, he says:

> *Out of the firey-fluidic state of the Earth, island-like formations emerged, and the beings upon them had an appearance similar to our present animals and plants. During the first half of the Lemurian epoch, the human being did not dwell directly on the Earth, but in the sphere above the Earth in a fine, rarefied body. Everything descends from higher spheres, including the human being, onto the continents currently undergoing densification.* (CW 109, June 10, 1909)

CHAPTER 19

The Animal-Human *at the* End *of the* Lemurian Epoch

Rudolf Steiner portrayed the Lemurian human being with distinct head, hands, and feet (Figure 114). Three animal beings (blue) seem to detach themselves from below the human being toward a violet-red, still warm and airy Earth while floating high above it. The pineal gland resides in the etheric body, indicated by the rosy glow in front of the head. The astral body was already internalized at this time in humans and animals. Rudolf Steiner first describes this image in general terms with the words:

> *The substances that later separated out of the common mass as liquid and solid were still air-like at that time. When the process of liquefaction began, the human body lived in the fluid element. Its locomotion was a kind of swimming-floating.* (CW 13, 1910)

In other places, he says:

> *The human being did not walk on their limbs like a human being of today. The etheric body was still outside the physical body and could transform the physical body because it was still much more flexible and supple than it is now. The human being could metamorphose and be reshaped by the will. The physical human being was as evolved as a mammal, except that mammals remained at this stage while the human being developed further. The branching off took place from a point where this original form had the possibility of ascending to the level of a human being on the one hand or descending on the other. The mammal's physical body hardened, while*

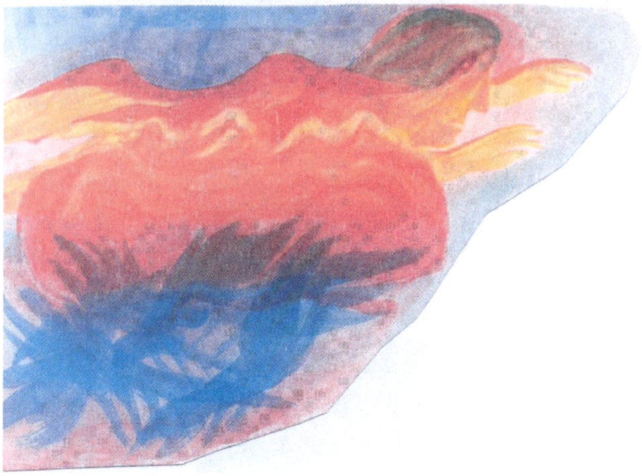

FIGURE 114 "Primeval Human - Primeval Animal." A watercolor painting by Rudolf Steiner for a eurythmy poster, 1924.

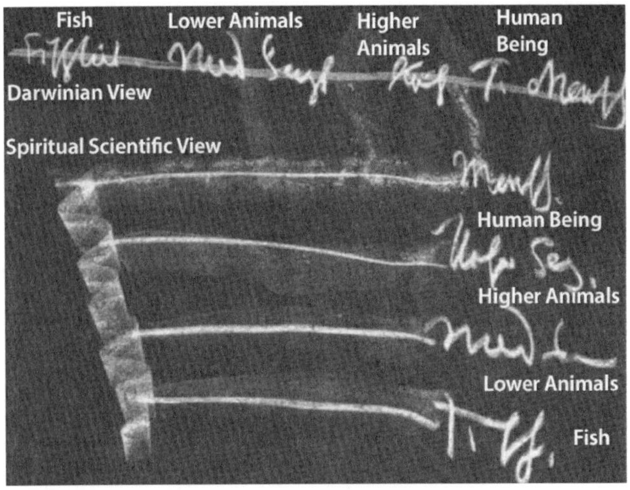

FIGURE 115 Ascending evolution from the viewpoint of Darwinism (upper horizontal line), as a continuous transformation of animal forms up to the human, and from the viewpoint of Spiritual Science (vertical line, below) as a stepwise ascending separation of individual groups (right) from a common spiritual origin (left).

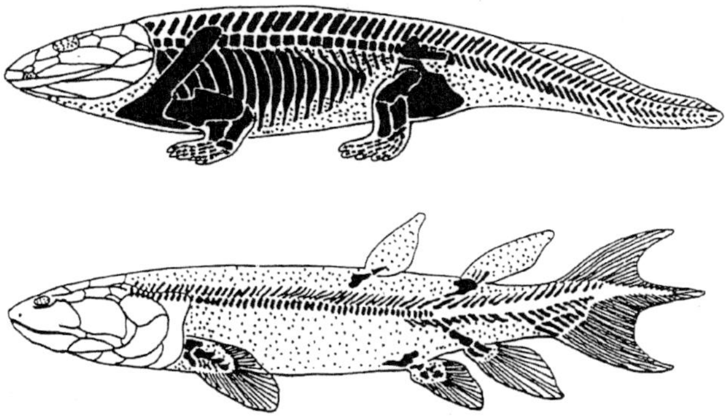

FIGURE 116 Fish (top) and an amphibian (bottom).

> *the human being could keep its body soft and malleable. Only in the Atlantean period did the other animal species branch off; in the oldest Atlantean period, certain lower mammals evolved, later certain higher mammals, and finally, today's apes branched off. We must not say that humans are descended from apes but that humans and apes are descended from one form.* (CW 100, June 22, 1907)

In the animal kingdom, the ego as a group soul causes the astral body to be internalized in the physical world. Animals react when strangers invade their habitat: dogs bark, cranes fly away, larks soar in circles. These reactions also evolved. Lower animals like fish, amphibians, and insects are silent. Slightly higher animals sound through their bodies: some frogs make a "Ribit Ribit" sound, while toads make a rapid series of popping notes, such as "du-du-du-du-du-...." Animals at a higher level react soulfully: blackbirds sing, crows caw, pigeons coo. And the mammals at an even higher level: dogs bark, cats meow, purr, and trill, lions roar, elephants trumpet, horses neigh.

The words of these reactions are taken from human experience. However, these animals express themselves while breathing in, while human beings speak or scream while breathing out! Try to bark like a dog when you breathe in! Animals often used for transportation, operations, and similar activities typically do not walk backward.

> *In animal groups with the highest levels of development, the ego, the next higher level of being, is already taking a formative role from the outside. Only in the human being at the highest stage of development can the ego be internalized. This is possible because humans have reached the evolutionary goal of all earlier stages. This makes higher spiritual development possible.* (CW 232, December 9, 1923)

Rudolf Steiner made a blackboard drawing that explains his view of human development as it has been presented to this point and is indicated on the left side by a vertical wavy line, in contrast to the Darwinian theory, which is shown above by a continuous horizontal line (Figure 115). Along the top horizontal line, the Darwinian sequence begins with fish, lower animals, and higher animals, with the human being as the end result. In contrast, spiritual

FIGURE 117 A hooded crow

FIGURE 118 A billy goat.

science begins with the human being removing the higher animals, lower animals, and the fish from its suprasensibly guided being. This idea is supported by the following figures. Figure 116 shows a fish as a backbone animal and above it an amphibian; in Figure 117, a bird, a hooded crow; and in Figure 118, a billy goat representing a higher mammal.

Rudolf Steiner says:

> *Let us look at the human head. The human head is, of course, a highly complex organ, but its most characteristic feature is this soft mass enclosed within a hard outer shell, and this basic feature can be found in a much more primitive state among the lower animals. We must look for an inwardly soft mass surrounded by a shell in the lower animals. And what is found is actually the simplest form for a human head. In crustaceans, the principle of the human head is in its most primitive form, and consequently, one can relate the human head to a shellfish. [...] If we observe the human chest system, the part of our body that is influenced primarily by the spine, we are led to higher animals, for example, fish. In a fish, the head is little more than a continuation of the spine despite its head being more differentiated. The fish is essentially a "backbone animal" that can be compared to the middle part of the human being. [...] If we look at still higher animals, the mammals, we must compare their high degree of development to the human metabolic-limb system. Such an insight gives us a proper view of evolution. Human development began with what finally emerged as the head, which happened during ancient times when the outer conditions of the Earth were entirely different from what they are today. The entire animal kingdom could be described as a human being spread out into all the manifold animal forms, or the human being is a synthesis of the entire animal kingdom.* (CW 303, January 1, 1922)

CHAPTER 20

The Old Atlantean Epoch

Rudolf Steiner says about the development from the Lemurian human being into the Atlantean:

> *During the Atlantean epoch of earthly evolution, human beings could not yet have existed in the form in which they now exist. If, for instance, you take a sea creature such as a jellyfish, which you can hardly distinguish from their surroundings, then you come to the material form in which the human being once was on Earth during the old Atlantean time, in which the Earth was still covered everywhere with a permanent, dense fog, in which the human being lived and therefore also had a completely different organic nature.* (CW 289, December 30, 1921)

Rudolf Steiner made a pastel drawing and a pencil sketch of the Atlantean epoch for the Goetheanum, shown in Figures 119 and 120. About this epoch, he says:

> *The present ego has evolved out of a group consciousness, or group ego. Now, there are four types of group souls. On one side, there is a likeness of a lion; on the other, a likeness of a bull; at the top, a likeness of an eagle; and in the middle, near the bottom, something that is already human-like. The Earth still had much softer matter, where air and water were not yet separated, where animals, plants, and all the beings that dwell on the Earth today formed in the surrounding fluid. When the mineral beings began to take on their present form, one could say that the human being was still a fine, airy being emerging from invisibility. They were surrounded on the outside by a shell when the animal was already in the flesh. The human being overcame these four animal types by descending to the physical plain.* (CW 104, June 19, June 21, and June 26, 1908)

The Old Moon was still together with the Earth into the Lemurian epoch, as is illustrated in Chapter 7 on the separation of planets. In Figure 121, Rudolf Steiner indicated the Moon's slow separation from the Earth. The sketch shows the basaltic with the thicker black dots and lines from left to right. The thinner dots indicate the airy, liquid environment of the life sphere from which the granite of the Earth's continents formed. There are numerous hypotheses in geology about the Moon's separation; for example, a giant meteorite could have hit the Pacific Ocean and caused the Moon to be ejected out on the other side, separating the continents. Rudolf Steiner says:

> *The separation of the Moon has not only taken place in the etheric, but the Moon as a physical world body has left the Earth and taken its own place in the planetary system.* (CW 233, December 26, 1923

> *The formations of the mineral kingdom remained solidified when the Moon separated from the Earth. Mighty catastrophes shook the Earth, and the basalt formations are the remnants of those purifying forces that shook the Earth's body when the Moon*

FIGURE 119 Pastel sketch of the Atlantian epoch by Rudolf Steiner for the Goetheanum. (Raske 1983)

separated. But this was also when the Earth's aqueous core became increasingly densified, and a firm mineral kernel gradually evolved. On the one side, the Earth grew denser through the Moon's departure; on the other, the upper parts gave off their granitic substances to the lower basaltic parts. (CW 106, September 7, 1908)

In Figure 122, Rudolf Steiner presented the transition of the human and animal from Lemuria above to Atlantis below. He says in simple words in his workmen lectures:

I will describe the state of the Earth, how on the Earth there was once a brownish-green, quite muddy surface, which sometimes became thick, then dissolved again. Above it was an air with all kinds of vapors and a heat like in an oven, above which clouds formed, with lightning and thunder. Bat-like animals lived in it and flew or swam around. They had very fine senses, like eagles, with which they could perceive the Sun and moonlight. If you went down into the muddy Earth, you would find huge animals, like dinosaurs, half waddling, half swimming, and waddling along (right). They had dolphin-like heads, whale fins, camel ribs, and huge shining eyes. And the Earth rotated much more slowly back then than it does today. But then came a time, shown below, when the Earth was already

FIGURE 120 Pencil Sketch of the Atlantian by Rudolf Steiner for the Goetheanum. (Raske 1983)

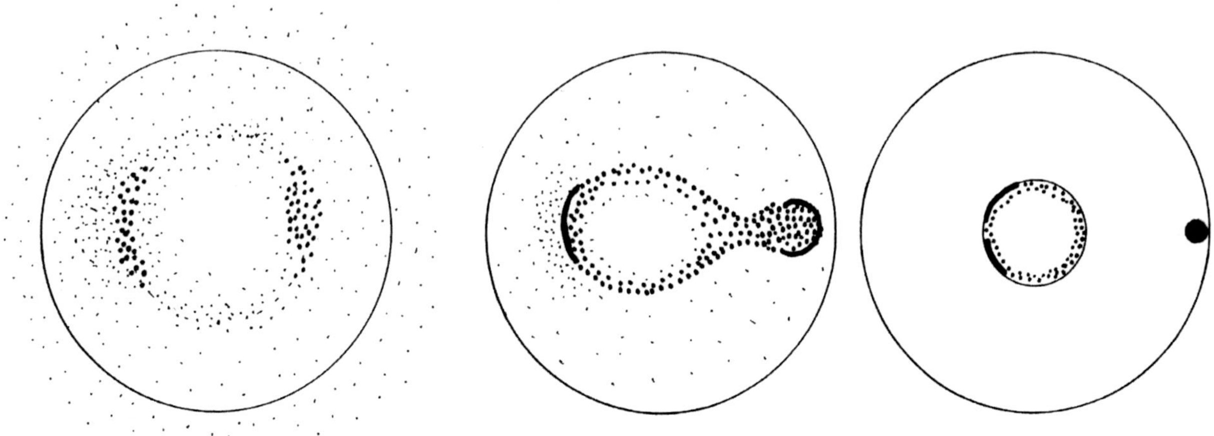

FIGURE 121 The slow separation of the Moon and the Earth during the Lemurian epoch, a process that likely began in the late Precambrian and continued into the Paleozoic. The basaltic with the thicker black dots and lines from left to right contrast with the thinner dots, indicating the airy, liquid environment of the life sphere from which the granite of the Earth's continents formed.

FIGURE 122 The change of condition from the Lemurian era (above) to the Atlantean (below). Above, reptiles half swam, half waddled in ooze with a fire mist above them with bat-like animals in flight. Below, the mud began to crystallize, and the watery air had cooled enough to support different life forms.

> *turning a little faster, the water-air had cooled down and let the harmful substances precipitate and merge with the Earth, although the entire Earth was still permeated by mists and vapors. The mud became denser and began to crystallize. And the fire-eaters became herbivores, eating the giant ferns. What they ingested internally became their external form. And the time came when the Sun could really affect the Earth. The dragonbirds gave rise to the later birds, of which wing prints have been found in softer limestone fossils. Originally, the Earth was alive, but the fact that it is dead today came about later. The Earth was once a giant animal, slowly turning in space, and human beings came and were able to inhabit it.* (CW 347, September 20, 1922, GAK 58, 25)

In another place, he says:

> *And whereas previously the human being existed as a finely organized being above the solid Earth in a kind of floating, hovering movement, he now had to descend from the Earth's circumference to parts of the Earth that were already more or less solidified.* (CW 13)

CHAPTER 21

The Mountain and Valley Cross in Europe *and* Micro-Earthquakes in the World

A mysterious pulse every 26 seconds from the Earth is a rhythmically occurring seismic movement or micro-earthquake whose epicenter is located in the Gulf of Guinea, off the west coast of Africa (See the online article in *Discover Magazine* "The Earth Is Pulsating Every 26 Seconds, and Seismologists Don't Agree Why"). Various phenomena can help us understand the Earth's pulse, including earthquakes and volcanoes, the directions in which the Earth's crust is formed, and the rhythms of the tides.

In Iceland, for example, glowing basalt has forced its way into glaciers that flow from left to right into the valley along a volcanic fissure (Figure 123). The ash from the rhythmic eruptions flies high into the air and then comes down again as light rain, blurring the picture. If you extend this fissure further to the south-southwest, about 30 km south of Iceland, you will find the basaltic island of Heimaey. Further in this direction, on November 14, 1963, at around 10 a.m., a volcano erupted from the sea floor from a 400-meter-long fissure and formed a new island called Surtsey, which is now 50 meters high at the top of the crater and is spewing lava. When I was invited to Iceland in 2000 to lecture,

FIGURE 123 A volcanic fissure in Iceland where glowing basalts have been forced onto glaciers. (photo Schutzbach 1976)

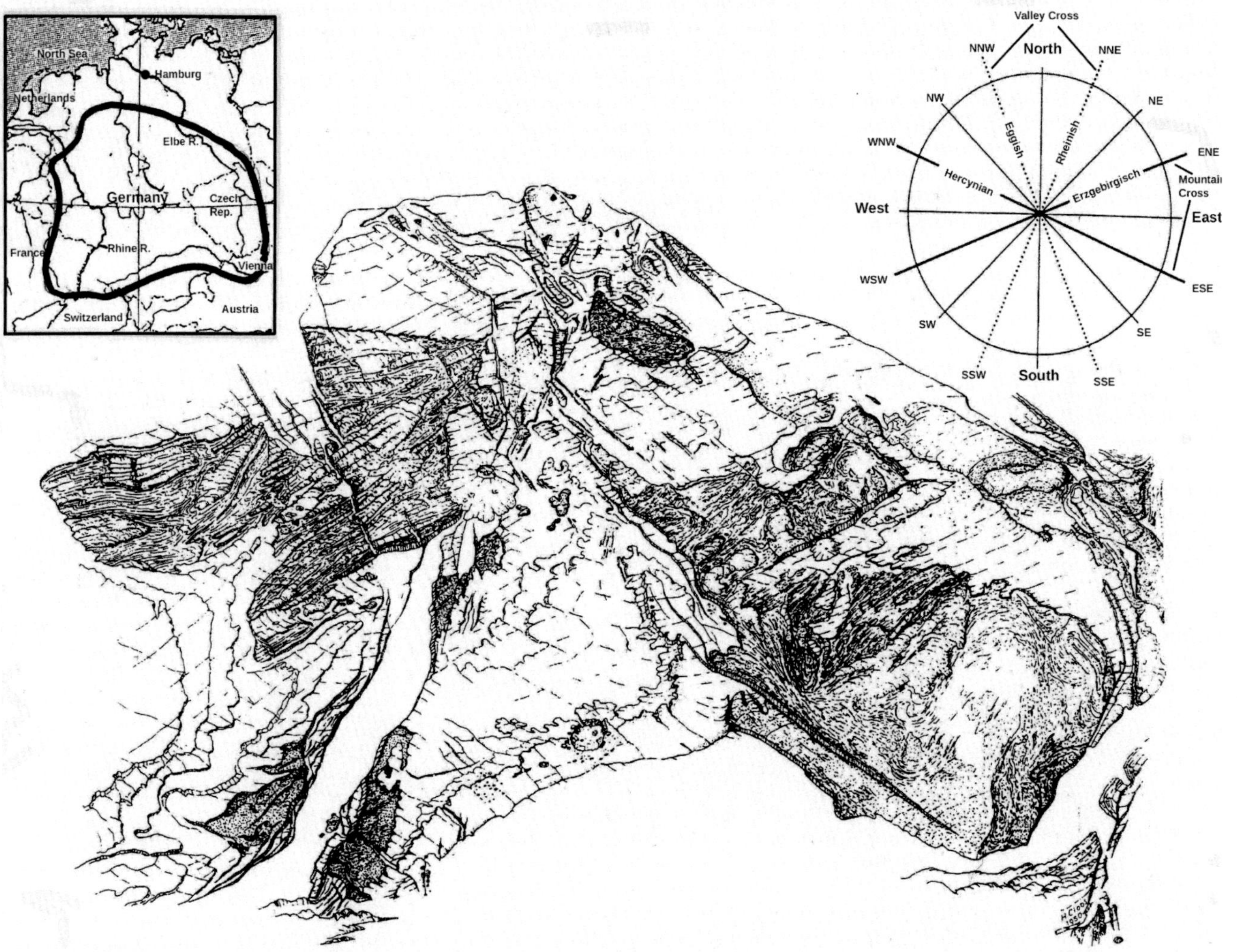

FIGURE 124 A tectonic map of Central Europe drawn by Hans Cloos. It presents the view of a geologist looking upon the landscape forms. Above right in this Figure are the four directions of the Mountain Cross and Valley Cross.

the country's highest volcano, Hekla, erupted ash and a little lava again. These islands lie on the submarine continuation of the Mid-Atlantic Ridge. Iceland is also famous for its geysers, which erupt regularly every hour and shoot water up to 70 meters into the air. The regularity of these eruptions is caused by heated and expanded groundwater. This process also occurs at Mount Vesuvius, which erupts in the Mediterranean. All eruptions are associated with tremors in the surrounding area.

The geological situation is very similar on the south coast of Nigeria, where the epicenter of the micro-earthquakes is located. As in Iceland, four volcanically formed basaltic islands lie in a line off the Atlantic coast in a south-southwest direction. They were formed in the Tertiary period when South America separated from Africa. Two of these islands are inhabited, and two are uninhabited. This direction is further extended by volcanic mountains to the north-northeast along the Central African Rift. The basalt mountains at Lake Chad and up to Tibesti were formed in the Tertiary period. The Jordan Valley, which contains the baptism site of Christ Jesus and the Dead Sea, also follows this direction. At 400 meters below normal sea level, it is the deepest depression on the Earth's surface!

This south-southwest to north-northeast direction of deep fractures from Africa continues

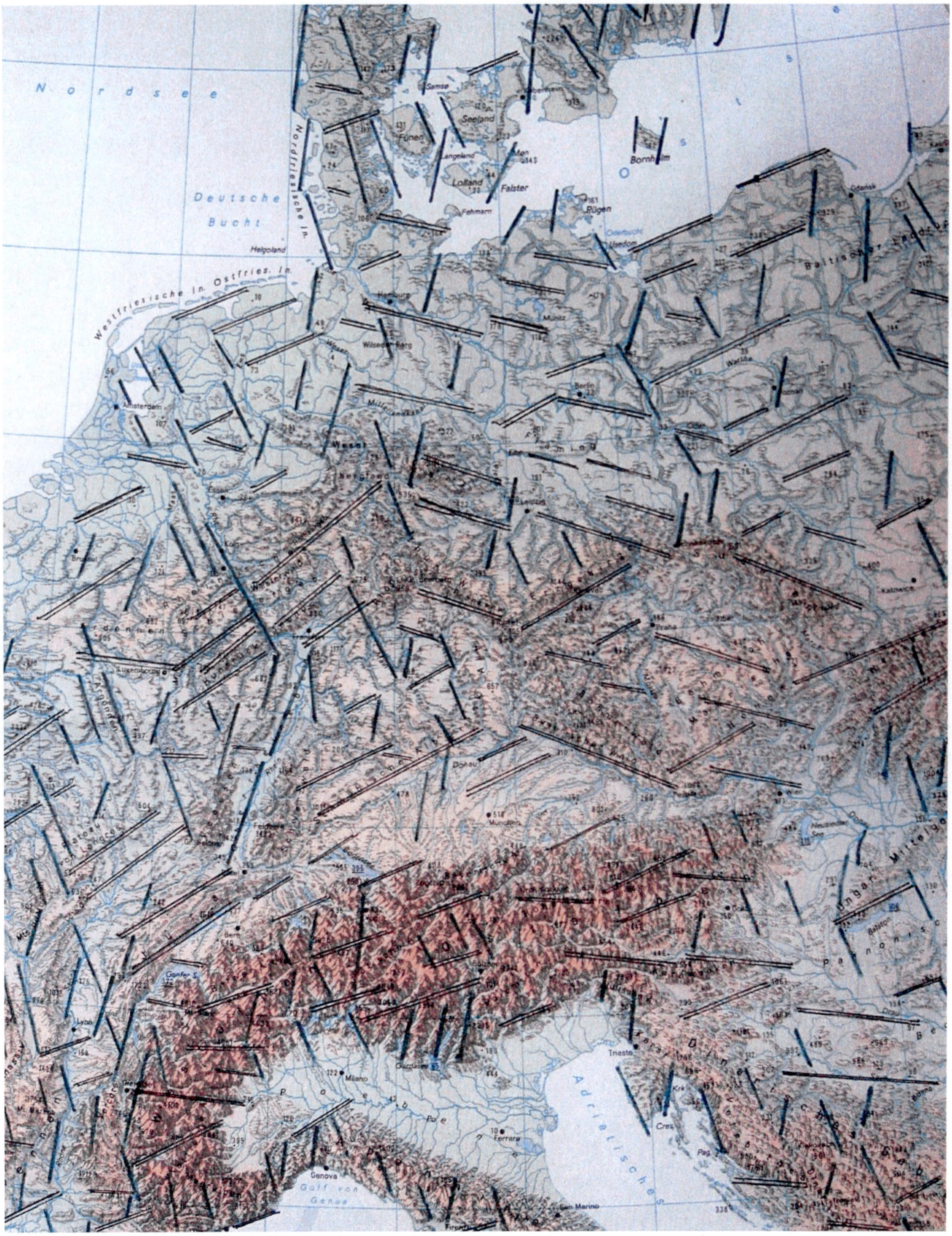

FIGURE 125 Map of central Europe in azimuthal projection. The Mountain Cross of the Earth refers to the two great high mountain ranges encompassing the Earth, oriented west-east in Eurasia and north-south in America. Double lines are associated with the mountain cross and bold single lines are associated with the valley cross referenced in Figure 124.

into Europe, with the Stromboli volcano north of Sicily and the Mount Vesuvius volcano near Naples. Further north is the area of central Europe shown on the maps in Figures 124 and 125. Figure 124 is a map of Central Europe drawn by geology professor Hans Cloos that shows how Central Europe's mountain and valley structure was shaped in four directions. Darker sections represent crystalline basement rocks at the surface, stratified rocks are lighter, and unconsolidated sediments are white. Above right in this figure are the four directions of the Mountain Cross and Valley Cross.

Two directions relate to the central mountain ranges. They were formed when the Alps were pressed from the south against the solid North European-Polish plate. These are the east-northeast to west-southwest Erzgebirge direction and the west-northwest to east-northeast Hercynian direction. These two directions are shown as double lines in Figure 125. In the North German Plain, both directions were created by the terminal moraines of the three ice ages, when the glaciers from Scandinavia reached their final form from the Baltic Sea coast.

In contrast, the other two directions were created by the separation of the North American continent from the European continent, forming the Valley Cross. One arm is the north-northwest to south-southeast Egge direction, named after the Egge mountains in the Rhineland. The north-northeast to south-southwest Rheinish direction is named after the middle Rhine Valley. These two directions lie along extensional surfaces extending through the 20 to 40-kilometer-thick granite continent down to the hot basaltic. This is why volcanoes can form along these two directions and always coincide with earthquakes.

When looking at the map in Figure 125, which is expanded over a wider area than the previous map, it is important to remember that the Earth's spherical shape, when projected on a rectilinear map, is always approximated over a longer distance. The azimuthal projection is used because all points on the map are at proportionally correct distances and directions from the center point. The grid lines in this map at full scale are 3.5 cm apart at the bottom and 2.5 cm at the top. Therefore, the valley and mountain directions can only be approximated over a greater distance!

Another phenomenon is the rhythm of the tides in the oceans, which are caused by the gravitational pull of the Moon. On the side facing the Moon, the tide rises; on the opposite side, facing away from the Moon, the tide falls. Imagine the Earth as a sphere turning under the mountains of water from west to east so that the tides alternate every six hours. These gravitational forces of the Moon also have a rhythmic effect on the volcanic fissures that penetrate the continents' groundwater zone and the mid-ocean ridges that trigger frequent seaquakes. These different rhythms of air and water characterize what we can understand as the life forces of the Earth's etheric body, as Rudolf Steiner calls it. This will be further illustrated in the concluding topics.

CHAPTER 22

The Precipitation *of the* Life Sphere *during the* Atlantian Epoch

The previous two maps of Central Europe are typical for many other landscapes and show how the varied hard surfaces of the rock crusts of all continents were shaped at rhythmic intervals. From the oldest Lemurian time onward, the water masses from the life sphere have been precipitated on this surface, resulting in our present-day landscapes. The seven figures in this chapter provide examples of this process. It is typical to see in the examples of calcareous-clayey sediments (Chapter 4) how each large or small landform builds a surface that rhythmically oscillates in infinite variation between hard and soft and thus recounts and tells the biography of how the temporal succession has merged into a spatial overlapping, as Hans Cloos perceived it!

This came about when the swirling air-water life sphere of the Lemurian epoch began to separate into air and water at the end of the Jurassic period. It separated the Earth's atmosphere upward to approximately 200 km into layers, including the Troposphere, in which the Earth's warmth usually allows a clear, fog-free view; the stratosphere, with its high cloud layers; and the ionosphere, containing the aurora borealis and where shooting stars light up. Today's atmosphere contains about 21% oxygen and 78% nitrogen, instead of the 3% oxygen found on Mars and Venus, where the atmosphere contains 95% carbon dioxide, while the Earth's air contains only 0.04%. The life sphere separated its water content downward. Part II describes how the continents occupied only one-third of the Earth's surface while separating from the planetary spheres; the other two-thirds of the cosmic body remained basaltic. In that area from the Mesozoic era onward, masses of water accumulated over the basaltic areas, especially during the Cenozoic era during the ice ages. An unimaginably large amount of ocean water filled the basins (363,500,000 cubic kilometers!), with an average sea depth of 3,700 meters; the deepest today is the Philippine Trench at 10,500 meters!

The ocean water is salty! It contains around 3.5% dissolved salts or 35 grams of predominantly sodium chloride (NaCl) per liter! In addition, there are some other substances with a concentration of one to three percent, for example, magnesium salts, which also cause a bitter taste. However, these salt contents vary in the seas and lakes; in the Dead Sea in Palestine, for example, it is 330 grams per liter! It is also the deepest depression on the Earth's surface, 400 meters below the normal water level of the oceans. Near the mouth of the Jordan River into the Dead Sea, Christ was conceived by God the Father in the body of Jesus of Nazareth and received on Earth by John the Baptist, as described in the Gospel of Luke.

At the very beginning of the Earth, during the recapitulation of Old Saturn in the Polarian epoch, spiritual beings of the highest hierarchies brought about the Earth's life sphere with its salt content. Rudolf Steiner explains the salts, rock salt (NaCl) and potassium salt (KCl), with the words:

> *That which is effective in forming the structure of any substance today, sodium or potassium, can be something that arose and worked a very long time ago, and perhaps was preserved. So, the original forms of our elements have been brought in from the cosmos, and the tendency to preserve these forms has remained even in the Moon period; the inner cosmos of the Earth is the reflection of the outer cosmos.* (CW 312, March 30, 1920)

Steiner's notebook about salt reads:

> *Thinking is extinguished in salt, thoughts solidifying; creating the basis for thought.* (Number NB 310)

In another place, he says:

> *It is characteristic of human beings in the present development on Earth that they participate in the mineral formation process. We can call it by its most characteristic feature, the desalination process, a tendency to transform salt formation into its opposite, and our humanity, above all our human thinking that goes beyond the animal, is actually based on this.* (CW 312, April 4, 1920)

What remains in the human being is about 70 to 100 grams of salt in the blood.

The original large continent with its granite-gneiss crust was almost completely hardened and crystallized at the end of the Paleozoic era. Chapter 5, about the gypsum and salt rocks, describes how the salts from the life sphere were precipitated in the Permian period at the end of the Paleozoic era. Salt deposits in later epochs are mostly rearrangements of salts from the Permian. After that, the water-air sphere remained undivided over most of the Earth until the end of the Mesozoic era. Rudolf Steiner remarks,

> *The previously empty ocean basins only filled up during the Ice Age. The air became purer and purer, not like today's air, but much purer. It only became a kind of water air in its later state, always permeated by dense fog vapors. The air was clearer earlier because it was warmer, and the substances it contained were more diffused. Later, it cooled off and became very foggy. This fog enveloped the entire Earth. Even under the Sun's influence, it did not completely lift.* (CW 347, September 20, 1922)

These processes can also be seen in the sense organs of the animals that developed during those epochs and can be observed, for example, in how bees or birds can orient themselves towards their nests in the evening or over large areas in the fall and spring! With the death of the life sphere, human beings also lost their previous habitat. After the vertebrates, humans in the Late Tertiary period were the last to form solid bones as they came onto the Earth's surface. This is described in greater detail in Part II.

It was not until the Cenozoic era that the air cooled to minus 50° C in the higher areas of the troposphere. Snow formed there which densified into glaciers in the polar regions. They became 2,000 to 3,000 meters thick during the ice ages and advanced to the northern continents in Central Europe, up to the low mountain ranges. It was only then that a circulation of water began, rising from the warmed surface of the Earth and falling down again as fresh water in the form of snow and rain. Today, 97.5% of

FIGURE 126 Loch Assynt in Scotland. Gravels and sands are still preserved on the peak plain in areas not incised by valleys.

the Earth's water is salt water, only 2.5% is freshwater, and two-thirds of this is bound up in the inland ice. Where the ice is thawing in the current warm epoch, most of the water flows into the oceans. Only a third, i.e., less than 1% of the Earth's water, ends up as fresh water in the ground, in rivers and lakes. Most of the world's groundwater also comes from the Ice Age. Fresh water is among Earth's most valuable natural resources because we need it to live!

Above the mountains that formed and folded mainly in the Tertiary period, the swirling water-air sphere formed a general flatness until the Ice Age, which always bounded the top of the mountain and is called the *Gipfelflur*, a peak or summit plain in the Alps, where the peaks and summits align at similar elevations. Gravels and sands are still preserved on the peak plains in areas not incised by valleys, which can be classified as the transition from the Mesozoic era to the Cenozoic era.

I found a good example of this concept at Loch Assynt in Scotland (Figure 126). The three island mountains in the background consist of 6,000-meter-thick solid quartzite that originated as sandstone. The granites and gneisses had already solidified and weathered into sands In the prehistoric Archean eon, as is the case everywhere where the Earth's bedrock lies at the surface. In the Proterozoic eon that came next, the then still complete continental blocks were eroded by the waters of the life sphere, scoured over large areas, and deposited worldwide as sands. These Archaean quartzites are usually red in color due to the oxidized magnetite of the basalt and can be found in many places on Earth. It was only later, from the Mesozoic era onwards, and especially during the ice ages, that the life sphere eroded all the overlying rocks worldwide and formed a first, uppermost plain. A remnant can still be seen on the hilltop on the right. Then, the deep, wide valleys between the three mountains were eroded down to the lower Archaean plain, where the roads run today, where the white quartzite rocks stand out as white spots, and where I took the photo. The steep, V-shaped valleys were carved during the ice ages. The Colorado River Canyon in Chapter 2, Figure 18, is in a similar landscape where the Grand Canyon exhibits cuts into these Archean quartzites.

Figure 127 shows an example of valley formation in the Central European Alps, where the Eisack Valley south of Bolzano is cut into the crystalline rocks of the Paleozoic era. The older planarity of the summit area is around 700 to 800 meters high and was formed by the end of the Tertiary period. Due to the further uplift of the Alps, the peak plain can reach over 2,500 meters. The cutting of the notch valleys began with the ice ages, which were first cut into the mountain and then partly covered with sand and gravel. The oldest high valleys were 10 to 30 kilometers wide and ran in the general direction of

FIGURE 127 Valley formation from the Central European Alps.

FIGURE 128 Melting glaciers on the edge of Greenland.

FIGURE 129 A glacier's surface with a crevasse.

today's rivers, including the middle or main terrace from which I took the picture at about the same altitude of 600 meters. The green forest on this side is part of a middle terrace formed during the second ice age.

On the northern continents, heavy precipitation formed the ice and the glaciers, as seen from an airplane (Figure 128), here showing melting glaciers on the edge of Greenland. Below them, the valleys become visible as they were also deeply incised by precipitation (as in Norway), not by ice erosion, as some hypotheses describe. The ice cannot erode solid rock but can only transport loose material into the foreland, which falls onto the glacier or has been loosened by frost weathering!

Figure 129 shows a glacier's surface and a crevasse formed when the ice contracts as it freezes. The snow falls at the top of the cirque where the glacier is formed, creating the lighter winter layers; the darker summer layers are formed when the ice thaws. If a person falls into such a crevasse, they are lost because their body heat melts the ice around them, trapping them unless they are rescued immediately by a helicopter. In this way, the inland ice glaciers, which are becoming ever more massive, such as those in Greenland, can be used to read the years, as mentioned in Chapter 11.

In the lowlands of Denmark, northern Germany, and Poland north of the low mountain ranges, and at the same time in southern Germany north of the Alps, the glaciers to the north of Scandinavia and to the south of the Alps have each advanced into the foothills. I have summarized the various landscape forms created by the glaciers in Figure 130. During a cold period, the ice pushes the loose, weathered material, stones (round), gravel, sand (dots), and clay (lines) in front of it and deposits it under the ice as a ground moraine. When a warm period occurs, the glacier stops and forms a terminal moraine in front of it, from which sand can be washed away toward the foothills. If such loose material falls onto the glacier, it can fall through crevasses and form an esker. When the glacier glides over previously formed valleys, and the ice there is cut off by the advancing glacier, glacial basin lakes remain after the melting. Similarly, rounded depressions and kettle lakes form under the ice. When the glacier finally melts, very shallow ice reservoirs remain, and the water can break through the terminal moraine and form deep channel lakes that flow further down into the glacial valleys. The valleys in central Germany were eroded during the same epochs. Each of the three ice ages formed additional terminal moraines, but these advanced less and less as the ice ages ended.

The living nature kingdoms live in continental areas that are often only 100 to 300 meters above the normal water level. The coastal shelf is seen in Figures 131 and 132. The continental slope drops steeply down to the ocean floor. Figure 131 shows an area off the east coast of North America adjacent to the Atlantic Ocean. Figure 132 shows an area off the west coast of the USA, the California longitudinal valley into the Pacific Ocean. Due to the enormous precipitation of the dying life sphere, submarine notch valleys were cut into these slopes around all continents during the Cenozoic era. They have main and side valleys, meanders, plunging and sliding slopes, and gravel terraces. These landscape forms indicate rivers carved them under atmospheric conditions and not underwater! This is now known worldwide, for example, from the Norwegian fjords, the St. Lawrence River, the Amazon, in Africa from Senegal, Congo, and the Zambezi; at the Spencer Gulf in South Australia, off the Indus, and off the mouths of the East Asian rivers into the Pacific (among others, *The Ocean Floor.* R. A. Scrutton, M. Talwani, 1982).

In summary, Rudolf Steiner characterizes the Ice Age period with the words:

> *In the middle of the Atlantean period, a new situation occurred. Similar to the end of Lemuria, it was again the development of human beings that led to the life forces of nature being betrayed and put into egoistic service. This led to the gradual destruction of the Atlantean region through air and water catastrophes. At that time, these storms gave the Earth a new face. The Earth's surface became solid and dry, which humans need to live. A solid skeleton was formed. The oldest bone finds are only from the end of the Tertiary period, i.e., from the middle of*

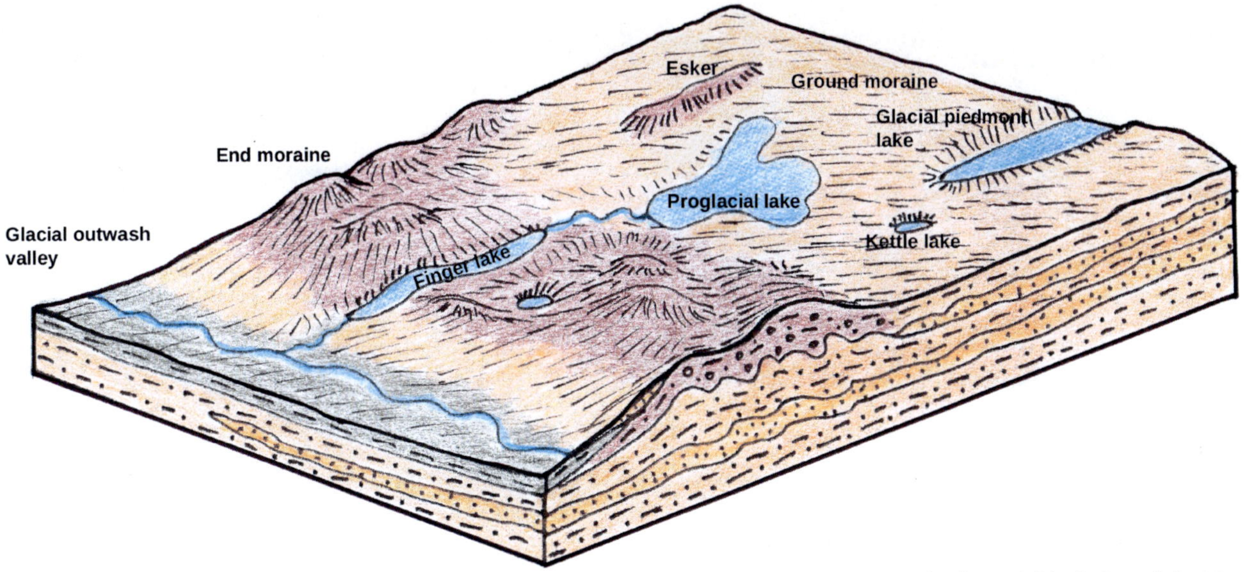

FIGURE 130 A schematic block view of glacial features in Northern Germany and the foothills of the Alps left by inland glaciers during the Ice Age.

FIGURE 131 An area off the east coast of the USA adjacent to the Atlantic Ocean.

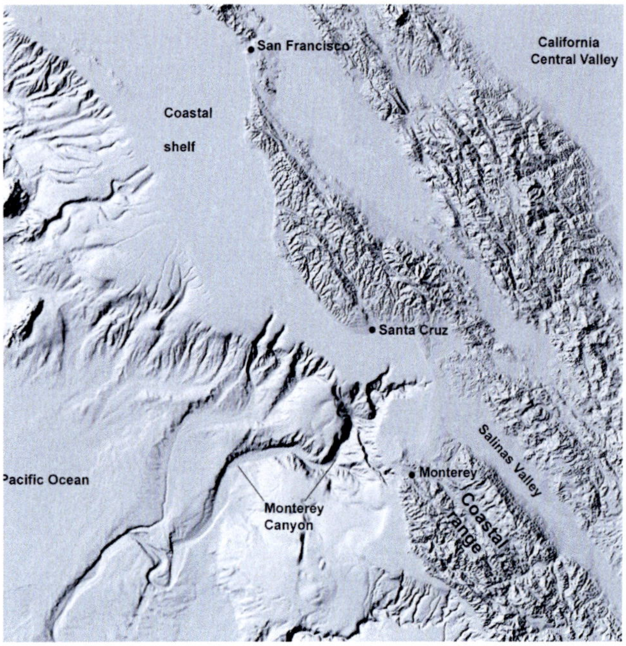

FIGURE 132 An area off the west coast of the USA: the California longitudinal valley into the Pacific Ocean (Heezen, Tharp, Berran 1969, detail).

the Atlantean epoch. The events on Earth are like an answer to this. Many of today's high mountain ranges were piled up. Only then did most of the water masses pour down from the atmosphere in enormous rhythms and fill up the ocean basins. The precipitation reshaped the previously uniform smooth surface of the entire Earth, creating landscapes; deep valleys were carved into the mountains, plains which today are only crossed by comparatively small rivers. Such landscapes are not known from the Earth's earlier history. The water masses filled the ocean basins, whose levels rose, and they also solidified in the glaciers of the Ice Age. (Parts summarized from two lectures by Rudolf Steiner in CW 99, June 4, 1907)

CHAPTER 23

The Emergence *of the* Nature Kingdoms *out of* Human Development

The questions of knowledge are ignited by granite! It was previously described how the continental rocks formed above the basalt are only found on Earth. However, the geology departments of universities cannot explain how this came about in the course of the Earth's history. The phenomena are researched from the mountains down to the smallest detail. This is the outstanding strength of the modern geosciences. The processes of formation are named with many terms that point to hypotheses that seem well-founded: *granitization, crystallization-differentiation, diatexis, anatexis, metatexis, metablastesis, palingenesis, auto-metasomatosis, regional metamorphism, plate tectonics.* When you question these, they are unclear pictures of conditions we do not know.

A general example of this is the word *magma.* Professor Wegmann, a professor of geology, vividly describes how the concept of magma came about:

> *When it was recognized that many crystalline rocks had penetrated other rigid rocks as melt flows, the material was called magma within this framework. Since the components were considered molten, high temperatures were applied to them. Because it was further assumed that all melt flows came from the depths, a rock cycle was considered.*

The sediments and *continental plates* would be submerged, melted in *currents within the Earth's mantle*, lifted into the mountains, eroded, and deposited as sediments. The formation of the original granite is explained by rocks that were only formed from the granite! This is meant seriously from a scientific point of view! Another professor of geology, K. R. Mehnert, at the Free University of Berlin, also describes many phenomena in his article: "50 Years of Granite Research." So now, after almost 100 years, all these questions are still open without a solution to how the first granite mass could have formed! The origin of the Earth's atmosphere, the water of the oceans, and the life of plants, animals, and humans also remain open.

The most extreme form of such hypotheses in today's science is the idea of the atom and molecule as if nothing else exists. For example, Professor H. J.

Rösler begins the description of crystal forms in his university textbook on mineralogy with the sentence:

> *Since we know that all matter is made up of atoms and molecules, we can conclude that these small building blocks produce this appearance through their regular arrangement.*

Another example comes from the Viennese professor of mineralogy, F. Raatz, who writes on these questions:

> *The principles of symmetry present us with the great riddle of what causes the formative power. It seems incomprehensible why the crystal surrounds itself with flat surfaces and exhibits those mysterious symmetrical properties.*

Modern atomic force microscopes make it possible to obtain much more detailed images of the inner structures of substances. They go much further than previous electron microscopy, namely to the extreme limit of visualization, where the possibility of resolution ultimately ends. This is the point at which it is said that atoms can be photographed. Computer images show almost infinite grids, leaving an unanswered question. They are interpreted as intersections of atoms and molecules, which in a vacuum consist of a nucleus that is orbited by colorless electrons at almost the speed of light on orbits that are shell-shaped at certain distances. Why don't they disappear into space? They would be kept at a distance and in equilibrium by electromagnetic forces. What are these forces? How can green petals turn into colored flowers in this way? Where do crystals of such different shapes come from, whose angles between the surfaces are substance-dependent and lawful? Or does the symmetry of crystals coincide with atomic concepts because, conversely, the symmetry of atomic grids and molecules is derived from them? Modern, materialistic natural science ends at the limits of knowledge.

Instead, the topics in this book can be used to show how the path of knowledge made possible by Rudolf Steiner enables phenomena visible to the senses and their suprasensible part to become a unified whole. In this way, one can learn to move clearly from visible phenomena of nature to the suprasensible developments of the world and back again to fully understand the phenomena. Rudolf Steiner confirms this with the words:

> *You will see that spiritual science links up everywhere with natural science in such a way that the latter on the pure basis of facts provides ample proof of what spiritual science as such has to offer. There are no grounds for contradictions between spiritual science and true natural science. [...] If you go to the sources, you can find, especially in geology, how geologists advance to a certain point and then place question marks. When you have arrived at these question marks, you can start from them by now looking at spiritual scientific research.* (CW 60, February 9, 1911)

Thus, some of Rudolf Steiner's blackboard drawings form the conclusion of this book. He has described the becoming of the Earth from various aspects. Concerning Figure 133, he says:

> *When we are fully acquainted with the animal and plant kingdoms, we can approach the crystallized mineral world. [...] But our whole perception is different. We experience something completely different when our gaze starts with the crystallized mineral than when we start from the plant world. We again become aware of "beingness" in the cosmos, and once more, we say to ourselves: what we see here below in the Earth as the crystallized mineral is brought about by a living, spiritual principle in the widths of the cosmos. But, as this works down (arrows from above), it is not reflected on or through the Earth (straight line). [...] It is as though the Earth were not there at all. [...] When we have concentrated on perceptions like this that arise from the crystallized mineral, [...] then below us is what at first is a frightening and terrible abyss, a void. [...] We feel we have lost the ground under our feet. [...] We must look through the Earth. Then we see everywhere currents and counter-currents, converging*

currents of cosmic energy from all directions. They meet below us, so from above, we have the streams for the plants – I have drawn it in yellow-green here – it goes down, the Earth offers resistance, and the plant grows out of it. But if we look at a stream for the mineral kingdom, everything is reflected in its own element. If you look at a quartz crystal, it is usually sitting on the ground. Still, it is only disturbed by the earthly; in reality, it is formed so that the spiritual element shoots together from all sides and is reflected geometrically exactly into each other. You see the quartz crystal, a six-sided prism, floating freely in the spiritual universe, forming completely on all sides. But now there are many crystal forms, a pyrite crystal that is perhaps dodecahedral; here we have a cube-shaped salt crystal. So, there are as many different cosmic streams as crystals in the Earth. We look into an immensity of worlds and say to ourselves: in the earthly realm, the deeds of many worlds meet, which live out themselves in mathematical-spatial form in the crystals. (CW 243, August 13, 1924)

Rudolf Steiner further elaborates on the process of quartz formation, which is typical of the mineral kingdom, by drawing a blackboard sketch (Figure 134).

The two white rays are representative of what acts from all sides out of the life sphere, showing what radiated in as finely distributed siliceous substances from the starry cosmos and what was already there in a certain sense in this albuminous substance, with a tendency to crystallize everywhere (white line). These imaginations had the form of gigantic, plant-like organisms, and out of that which was "imagined" into the Earth in this way, there developed later, through absorption of the atmospheric substance — the plant; everything that is of a plant-like nature. At first, it was in the environment of the Earth (white dots), in a volatile, fluid form; only later did it sink down into the soil and become what is known to us as the plant. In addition to the silicic acid, another finely distributed substance was embedded in this albuminous atmosphere – chalky substances (red). Again, out of the chalky substances, under the influence of the congealing albumen, there arose the animal kingdom. The human being felt united with all of this (circumference), and in the beginning of time, was one with the whole Earth. The human being lived in what was forming on the Earth as plants through suprasensible images and lived in what was forming on the Earth as animal. And the human being felt spread out over the whole Earth, one with the Earth. (CW 233, December 26, 1923)

The Earth was extended as far as the Moon sphere in the Lemurian epoch (Figure 135).

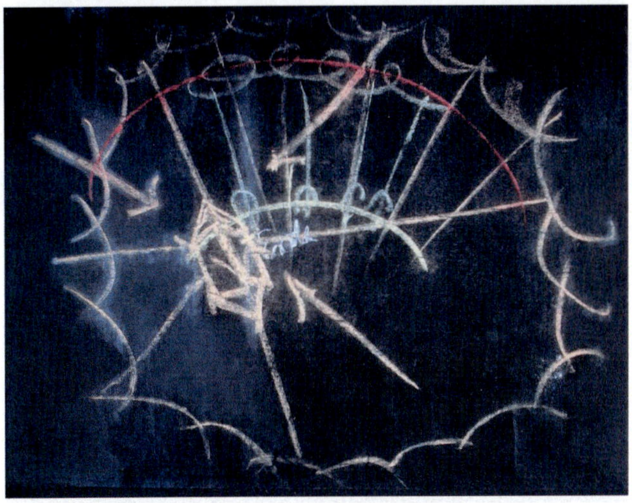

FIGURE 133 Blackboard drawing by Rudolf Steiner, August 13, 1924, CW 243.

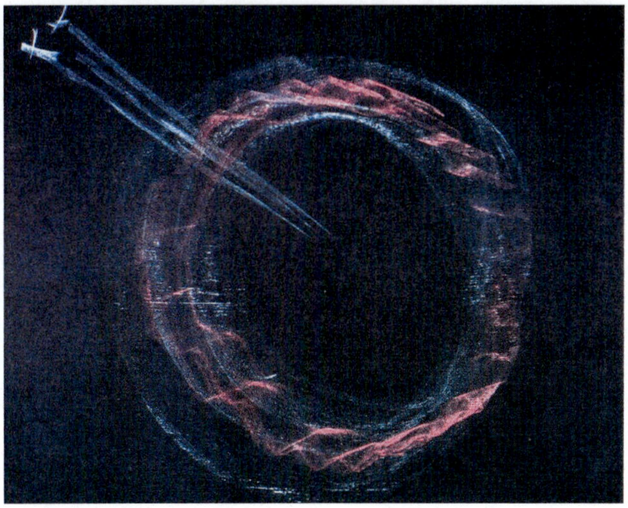

FIGURE 134 Blackboard drawing by Rudolf Steiner, December 26, 1923, CW 233.

> *Today, we speak of the earthly element, the watery element, the airy element, the fiery or warmth element, and then about the ethers. This drawing represents the solid Earth (white core), the watery liquid Earth (light red), the air (blue), and the whole is permeated and surrounded by fire (deep red). But for Aristotle, the Earth extends up to the Moon. And from the cosmos inward, from the stars, from the zodiac inward to the Moon, no longer into the earthly realm, the remaining ether extends spatially and cosmically (white outside). The ethers beyond earthly warmth — the light ether, the chemical ether, and the life ether — were also once connected to the Earth and went into the Earth. But the ether withdrew from the Earth when the Moon withdrew in the old evolution. [...] When Spring approaches, the elemental beings of the Moon bring the ethers from the Moon again to the beings that are the plants, animals, and humans. In this way, the Moon is the shaper and molder of beings.* (CW 233, December 27, 1923)

The next blackboard drawing (Figure 136), also by Rudolf Steiner, is another cross-section of the Earth that describes the evolution of the living Earth in three phases. The first phase, the Lemurian, is from the Proterozoic eon to the end of the Paleozoic era, with the separation of the Moon and salt precipitation; the second is the Mesozoic era, and the third, the Atlantean, the Cenozoic era with its ice ages.

> *The human being was already united with the Earth but still entirely spiritual (yellow). Something entered the air sphere that was incredibly fluid, similar to the later albumen. The cosmic forces acted on some parts of the albumen so that it coagulated and formed the siliceous (white), and from the granitic, the sulfuric, and the metallic, the ores were formed. In it was life in plant forms (green). To this was added what you find today of animals and limestone, but it was like a continuous upward and downward movement of the limestone (red).*
>
> *In the second phase, the whole atmosphere was transformed, in which the human being brought forth somewhat more solidified creatures, which appear as the descendants of the earlier plants and animals that geology finds in impressions. By forming the whole of your being from heaven into the Earth, you have separated it from yourself, made it the foundation, and given it over to pure mineral life so that a new life could develop upon it. This forms the ground on which we walk. It contains what the kingdoms of living organisms could not use, which they could thus transform to walk on. It is an ascending current that developed a higher kind of plant and animal life and another current that continued the decomposition by which the sediments were formed over the crystalline granites and gneisses.*
>
> *Finally, a third phase of creation began. As the Earth approached that time, it was gripped by all kinds of destructive processes, which had the effect of a spiritual soul working on the material. The processes become less and less full of life and more and more chemical and mechanical, so in our time, we only have water effects, which continue the old processes of destruction. And there, in the Atlantean period, plants and animals developed into life forms that were already similar to those in our time. This is when the Earth is shaped into what it is now. All development is based on the fact that something independent is first separated from the life of the surrounding environment; upon this, the environment is imprinted as if by reflection, and then this separated being continues to develop independently.* (CW 233, December 28, 1923).

Three more images that characterize the descent of the nature kingdoms from the human being and Christ bring us closer to the conclusion of this book. In Figure 137, the circles indicate the essential members of humans, animals, and plants. On a dark background toward the bottom is a white arc designating the Earth, and above the white arc is the physical world where the human, animal and plant have been identified. The next layer is the astral or elemental world, shown in yellow. Above that, where it is again dark, is the lower spiritual world. Beyond this is the upper spiritual world that works in from the cosmos. The word "plant" was written diagonally because the blackboard ended. Because of this issue, the mineral kingdom does not fit on the blackboard drawing. For completeness and clarity, I added the mineral kingdom to a modified blackboard drawing, as seen on the right side of Figure 138. This modification results in another circle for the upper spiritual world.

Figure 138 gave rise to the diagram of the descent of the nature kingdoms (Figure 139), which illustrates how the essential elements of minerals, plants, and animals have been separated from the ego of the human being in the course of the Earth's history. The higher development of human beings is based on the fact that the essential members of their being incarnated one after the other. This descent is shown by the rose-pink circles. Our ego, in turn, originates from Christ, who works in the entire circumference of the Earth, which is indicated by the red arrows. Spiritual beings work respectively from the world spheres indicated on the left into the beings of the nature kingdoms and the world spheres indicated on the right, down to the physical world. Just as the rose-pink circles denote the ego entity, the dark yellow circles point to the astral body, the green circles to the etheric body, and the blue circles to the physical body.

Human beings have their four bodies in the physical world on Earth; in animals, the physical body and etheric body are also in the physical world, but the astral body is in the soul world, which is led by the animal groups known as herds, flocks, or species. In the case of plants, the physical body and the etheric body are on Earth, which causes growth, and the astral body is in the soul world, which causes color and scent. With the forces of the lower spiritual world, the seed is created in the flowers, which in the following years causes them to spread and additional plants to grow. The mineral forms its mathematically exact shapes in the upper spiritual world, the rocks in the lower spiritual world, the ore substances radiate in from the planetary sphere, and the "I" causes shapes in the physical world on Earth, which is

FIGURE 135 Blackboard drawing by Rudolf Steiner, CW 233, December 27, 1923.

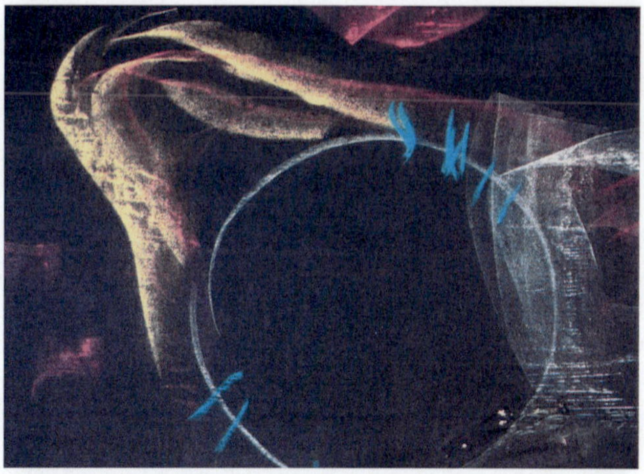

FIGURE 136 Blackboard drawing by Rudolf Steiner, December 28, 1923.

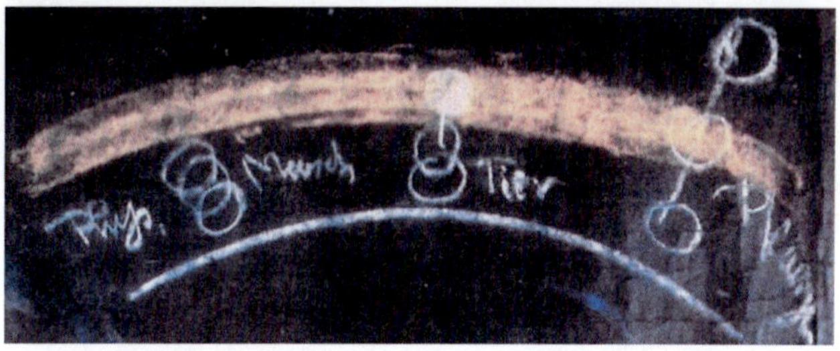

FIGURE 137 Blackboard Drawing by Rudolf Steiner, CW 243, August 11, 1924)

FIGURE 138 Blackboard Drawing by Rudolf Steiner, modified by the author to include the mineral world.

different for every crystal! Each picture in this book shows an example of this infinite variety and thus provides a good overview.

Rudolf Steiner characterizes two directions of cognition that must be given special consideration today:

> *After the Mystery of Golgotha, we have two currents of human development, the one which carries the impulse of Christ within it and leads gradually upwards into the spiritual heights, and the other which is a continuation of the descent and leads into material life, into materialism. Today, human beings must look without prejudice and impartially at everything that spiritual science can tell us so that they can realize that the soul needs inner progress in the sense of spirituality to the outer world, which is necessarily becoming more and more materialistic. In the spiritual current, we must learn to question. In the materialistic current, people are led away from questioning. It is considered freedom and of increased personal value if you can make your own judgment everywhere. We only learn to ask questions if we do not judge and if we can put ourselves in an expectant mood so that events in life may reveal themselves, if we can wait for what is to flow out of the suprasensible realms of existence. Try to realize through meditation the difference between approaching the spiritual realms of life with judgments or with questions. Then, you have to experience inwardly that there is a radical difference between the two. The spiritual current of thought will only flourish if we learn to understand the difference between questions and judgments. Certainly, we must make judgments concerning the external circumstances of life. Our*

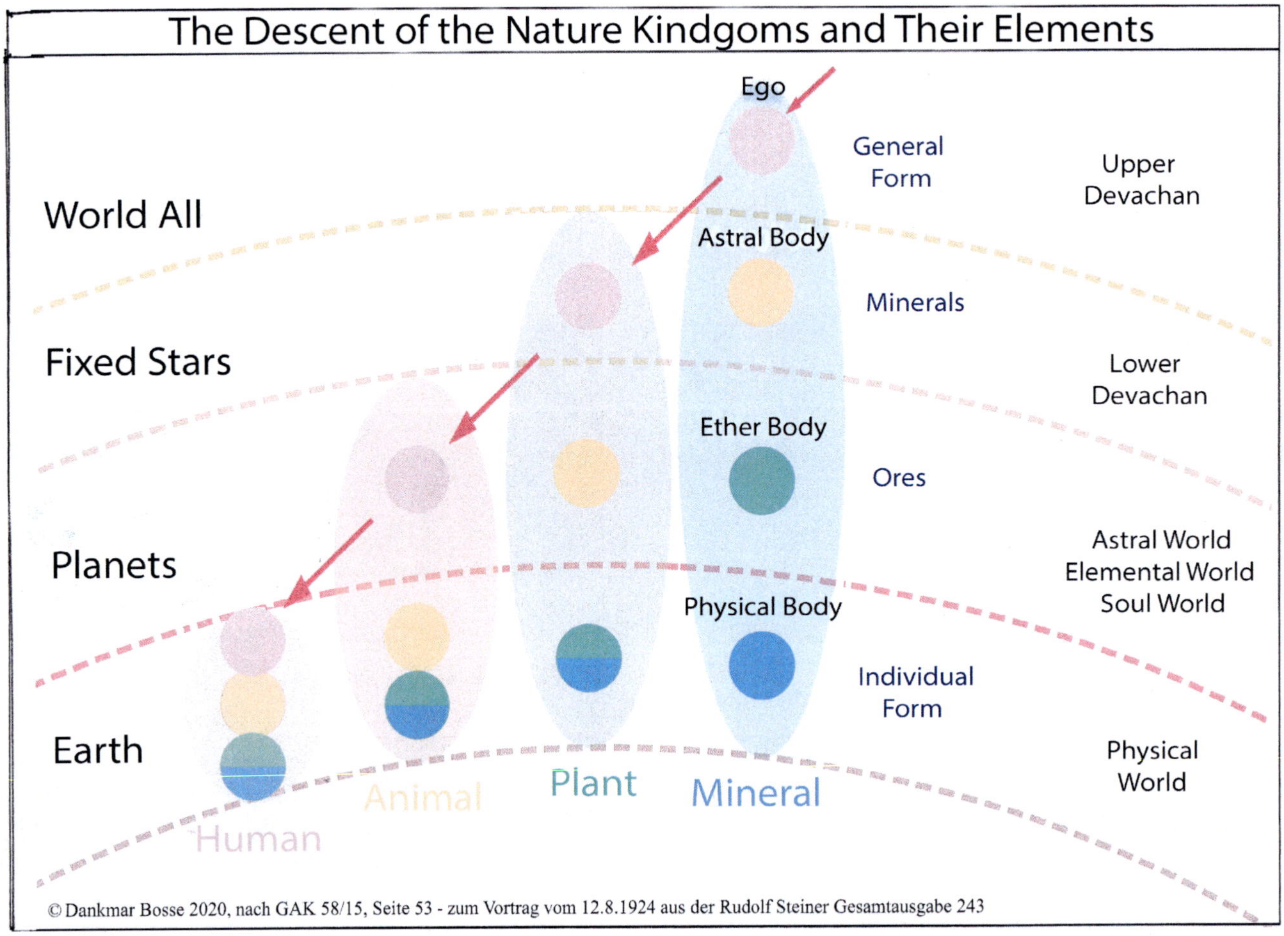

FIGURE 139 The Descent of the Nature Kingdoms and Their Elements.

> *spiritual movement will go further through everything by which this questioning mood is recognized and promoted in a larger part of humanity.* (CW 148, January 6, 1914)

I have gained the possibility for such a meditation exercise to progress further along the boundaries of knowledge from suggestions by Goethe and Rudolf Steiner. The Goethean method of cognition has five stages of practice.

STEP ONE

The first step on this path is thoroughly ***observing phenomena*** without presupposing theories or ideas. This step is also necessary in natural science. It cannot be about the thoughtless experience of pure phenomena, because observations cannot do without concepts. This is the great gain that the materialistic sciences have brought, that the phenomena are known exactly. From there, inner questions about nature can open up and direct the field of observation.

STEP TWO

In the second step, the ***sequence of phenomena*** is important, either in the order in which it is given by nature or how a sequence is logically constructed by the experimenter. These orderings can be obtained sequentially, rhythmically, or in spatial juxtaposition. For example, in the case of rocks, the sequence of superimposed layers with their animal fossils; for plants, it would be the sequence of leaves; for animals, their life cycle. This makes it possible to read the time sequences from the relationships of the adjacent layers and thus recognize the development processes. The most important thing here is to pay attention to the transitions; what distinguishes one phenomenon from the adjacent one or unites with it. Goethe recommends that the best ordering is the one by which the details become a great phenomenon, a kind of biography that can show us something of the essence.

STEP THREE

In the third step, we can obtain what Goethe calls:

> *a **pure phenomenon**, a basic or archetypal phenomenon ultimately resulting from all experiences and trials. To represent it, the human mind determines the variable and excludes the accidental. This leads to a vivid overview, a general concept into which the particular case can be integrated. I had to stick to my old way, which compels me to view all natural phenomena in a certain sequence of development and to attentively follow the transitions forward and backward. In this way, I arrived from within myself at a living overview, from which a concept is formed to meet the idea in an ascending line.*

This is how he formulated it in the "Metamorphosis of Plants."

> *Forward and backward, the plant is always only a leaf.*

STEP FOUR

This brings us to the fourth step that Rudolf Steiner calls the ***saturated* concept**. The idea can reveal itself from the other side, from the world of the spirit, as something essential. Goethe writes:

> *Every new object, well looked at, unlocks a new organ in us.*

He calls it the ability to *see with the eye of the spirit*. By this, he means perceiving an idea not as a random idea, concept, or theory but as a *spiritual reality* as an inner experience.

STEP FIVE

The result is a fifth step, with which one must look back to the beginning of the path. It is important to compare the idea gained with the observations at the beginning, i.e., to ***test it against perception*** and to enter into a dialogue with it.

> *The phenomena must not be looked at with the eyes of the body but with the eyes of the spirit; one must become accustomed to seeing simultaneously with the eyes of the body and the spirit.*

The Goethean path of practice can lead to the

> *elements of the objects, the views, entering into the faculty of thought and being penetrated by it most intimately.*

In this way, a free answer appears to the inner question, which is experienced clearly, not dreamily, not as an imagined conception. It can arise in the questioning gaze, in the awake observation of a phenomenon or a landscape, like a *real* idea, as the possibility of an answer, which is experienced like a weighing or proving, like a spiritual conversation between the phenomenon and thinking. For example, this method has proven itself to me and other friends over decades of use.

Other people have also described a similar experience. The professor of geology Hans Cloos, wrote in his autobiography *Conversation with the Earth*:

> *Is not the beauty of the landscape the unconscious insight into the inner order of the Earth, into the rhythm of repetitions, into the harmony of planes and lines, felt by a being that is itself part of the same nature, and thus inwardly subject to the same orders it encounters outside itself, and the strings of our soul excited to vibrate with the music of the Earth?*

Author Biography

DANKMAR BOSSE was born in Weimar, Germany in 1940. He studied geology, paleontology, and mineralogy at the Freiberg University of Mining and Technology in Saxony (East Germany) and worked for twenty-one years at the Central Geological Institute in East Berlin, mainly on regional geology and middle European tectonics. Since then, he has been independently active with seminars, lectures, and anthroposophical research work in all earth-scientific fields and numerous research excursions.

He is the editor of the book *J. W. Goethe: Die Metamorphose des Granits* (*J. W. Goethe: The Metamorphoses of Granite*), an edited collection of Goethe's geologic writings, published by Verlag Freies Geistesleben in 1994. In 2015, his comprehensive two-book set on mineralogy from a natural-science/anthroposophic viewpoint *Die Evolution der Minerale zwischen Kosmos und Erde* (*The Evolution of the Minerals between Cosmos and Earth, Sketch of a Mineralogy and Crystallography of the Living Earth*), was published by the Verlag des Ita Wegman Instituts, Stuttgart (untranslated). In 2019, *The Mutual Evolution of Earth and Humanity: Sketch of a Geology and Paleontology of the Living Earth* was published. It was followed by additional essays on the same subject in *Essays from Earth's History* in 2021.

Partial List *of* Sources

RUDOLF STEINER: CW # or GA # refers to the Rudolf Steiner Complete Edition; Writings and Lectures, Rudolf Steiner Verlag, Dornach, Switzerland. They are given with the date of the lecture. The words quoted are written in the present tense because they retain their validity: Rudolf Steiner "says" rather than "he said," as if it could be said differently later! I have taken the liberty of shortening the texts from the lecture style to the reading mode without changing the statement's content!

JOHANN WOLFGANG VON GOETHE: The first editor of Goethe's work was Rudolf Steiner, who strove to create a text for the Weimar Sophien Edition organized by content and thus readable. This proved impossible because the philological standard applied was to produce a source edition that was a literal translation of the original. Because the original goal was not achieved, the plan for the Leopoldina Edition emerged in the 1930s. A text that was readable in terms of content for geology and mineralogy was only achieved in 1960 by Helmut Hölder in the Cotta edition.

DANKMAR BOSSE: Unless otherwise noted, the author of the images and text in this book is Dankmar Bosse, including other titles listed in his name below.

1. Dankmar Bosse, *Die Evolution der Minerale zwischen Kosmos und Erde: Entwurf einer Mineralogie und Kristallografie der lebendigen Erde*, (*The Evolution of Minerals between Cosmos and Earth: Outline of a Mineralogy and Crystallography of the Living Earth*), Verlag des Ita Wegman Instituts, 2015. Untranslated.
2. Dankmar Bosse, *Die Lebenssphäre der Erde*, (*The Life Sphere of the Earth*), Verlag des Ita Wegman Instituts, 2012, ISBN: 978-3-905919-37-0. Untranslated.
3. Dankmar Bosse, *The Mutual Evolution of Earth and Humanity: Sketch of a Geology and Paleontology of the Living Earth*, Lindisfarne Books, 2019. Translated by Frank Fawcett.
4. Dankmar Bosse, *Essays from Earth's History: Additional Studies Concerning the Mutual Evolution of Earth and Humanity*, Formative Forces Publishing, 2021. Translated and published by Jannebeth Röell and James Lee.
5. Rudolf Steiner, *Blackboard Drawings 1919-1924*, Rudolf Steiner Press, 2003.
6. Rudolf Steiner Archive in the USA: https://rsarchive.org This site contains the text of many lectures, works, and articles by Rudolf Steiner that have been translated into English and information on the Goethean observation method.
7. Rudolf Steiner Archiv in Dornach: https://www.steinerverlag.com/de/CW-online This site is a commercial archive assembling all of the German language lectures, books, notes, and other source information concerning Rudolf Steiner.
8. *The Complete Works of Johann Wolfgang von Goethe* in Ten Volumes: https://archive.org/details/completeworksofj05goet
9. Mindat.org is the world's largest open database of minerals, rocks, meteorites, and the localities they come from, with a strong listing of minerals, their chemistry, synonyms of mineral names, mines, etc. https://www.mindat.org
10. R. van Romunde, *About Formative Forces in the Plant World* with original watercolor drawings by Elly van Hardeveld. Translated and published by Jannebeth Röell, 2001.
11. Walther Cloos, *Gemstones and Earth's History*, Formative Forces Publishing, 2021. Translated and published by Jannebeth Röell and James Lee.

OTHER AUTHORS

Amann, G.: *Bäume und Sträucher des Waldes*, Munich, 1954.

Bentley, W.A.: Humphreys, *W.A. Snow Crystals*, New York, 1931, reprint 1962.

Brandt, M.: *How old is humanity?* Baiersbronn, 2006.

Cloos, Hans: Drawing published posthumously in 1958, Freiburg i.Br., and Stuttgart, 1970. Chapter 21, Figure 124.

Cloos, Hans: *Conversation with the Earth*, New York, Alfred A Knopf, 1967. Translated from the German by E. B. Garside.

Cloos, Hans: *Gespräch mit der Erde*, Leipzig 1947, ISBN: 978-3-8262-3063-9, text in Chapter 23.

Goldschmidt, V.: *On Harmony and Complication*, Berlin, 1901.

Goethe, J.W.: *Leopoldina-Ausgabe I*, 9, S. 114-115, Text: Nacharbeiten und Sammlungen, Weimar, 1954. Bild-Tafel: Goethes Sammlung zur Botanik: Anna-Amalia-Bibliothek Weimar: Corpus der Goethezeichnungen, Bd. V B, S. 116-120.

Mehnert, K.R.: "50 Years of Granite Research," *Geol. Rundschau*, 76/1, p. 1, for cited text in Chapter 23, Mond, Stuttgart, 1987.

Newell, H.E. (ed.): "Satellite Explores Earth and Moon," *Luna Orbiter* 4, p. 94, Frankfurt a.M., 1969. Chapter 7, Figure 61, Moon.

Owen, H.G.: *Atlas of Continental Displacement*, Continents, Redrawn for Images: Granite, with deposit signatures according to Implications of Continental Drift II, p.1057, 1973, Cambridge, 1983. Cover Map of Pangaea and Chapter 10, Figure 80.

Raatz, F., Köhler, A.: *Bau und Bildung der Kristalle*, p. 25; Vienna, 1953.

Raske, H.: *Das Farbenwort. Rudolf Steiners Malerei im ersten Goetheanum*, Stuttgart, 1983.

Rösler, H.J.: *Lehrbuch der Mineralogie*, p. 257, for Widmannststättensche Figure in Chapter 7, Figure 60. Leipzig, 1981.

Said, R.: *The geological evolution of the River Nile*, Berlin, 1981

Scrutton, R.A. & Talwani, M.: *The Ocean Floor*, New York, 1982

Schutzbach, W.: *Feuereinsel am Polarkreis*, Bonn, 1985 for Chapter 9, Figure 123, Iceland.

Wimmenauer, W.: *Petrographie der magmatischen metamorphen Gesteine*, Stuttgart, (1985); and the essay (1994): "Der Granit - 200 Jahre nach Goethe." Tycho de Brahe-Jb., S. 116.

Winkler-Hermaden, A.: Geologisches Kräftespiel und Landformung. Vienna, 1957.

Yardley, Bruce, W.D.: *Introduction to the Petrography of Metamorphic Rocks*, Stuttgart 1997, pp. 196-224.

Time Period according to Rudolf Steiner	Geological Time Period			Layers	
Post-Atlantean	**Cenozoic**	Quaternary	Holocene (Post-Ice-Age)		**Unconsolidated sediments layer**
Second Half of Atlantis			Pleistocene		
Middle Atlantis		Upper Tertiary	Pliocene Miocene		
First Half of Atlantis		Lower	Oligocene Eocene Paleocene		
Destruction of Lemuria	**Mesozoic**	Cretaceous	Upper Lower		**Stratified rock layer**
Late Lemuria		Jurassic			
		Triassic		↑ Secondary rocks *Mainly* Primary rocks ↓	
Middle Lemuria	**Paleozoic**	Permian			Transitional layer:
		Carboniferous			**Crystalline basement layer:**
		Devonian			
		Silurian			Schistose layer
		Ordovician			
		Cambrian			
Early Lemuria	**Proterozoic**	Precambrian			Crystalline layer
	Archaen			Stromatolites Bacteria Archaebacteria ↑ Bacteria-rock branch ↗	Primary rocks Upper Earth's crust Lower
Hyperborea				↑ Mineral-plants ↑ Mineral-plant-animals ↑	
Polaris				Human ancestors →	Plants Animals → Humans

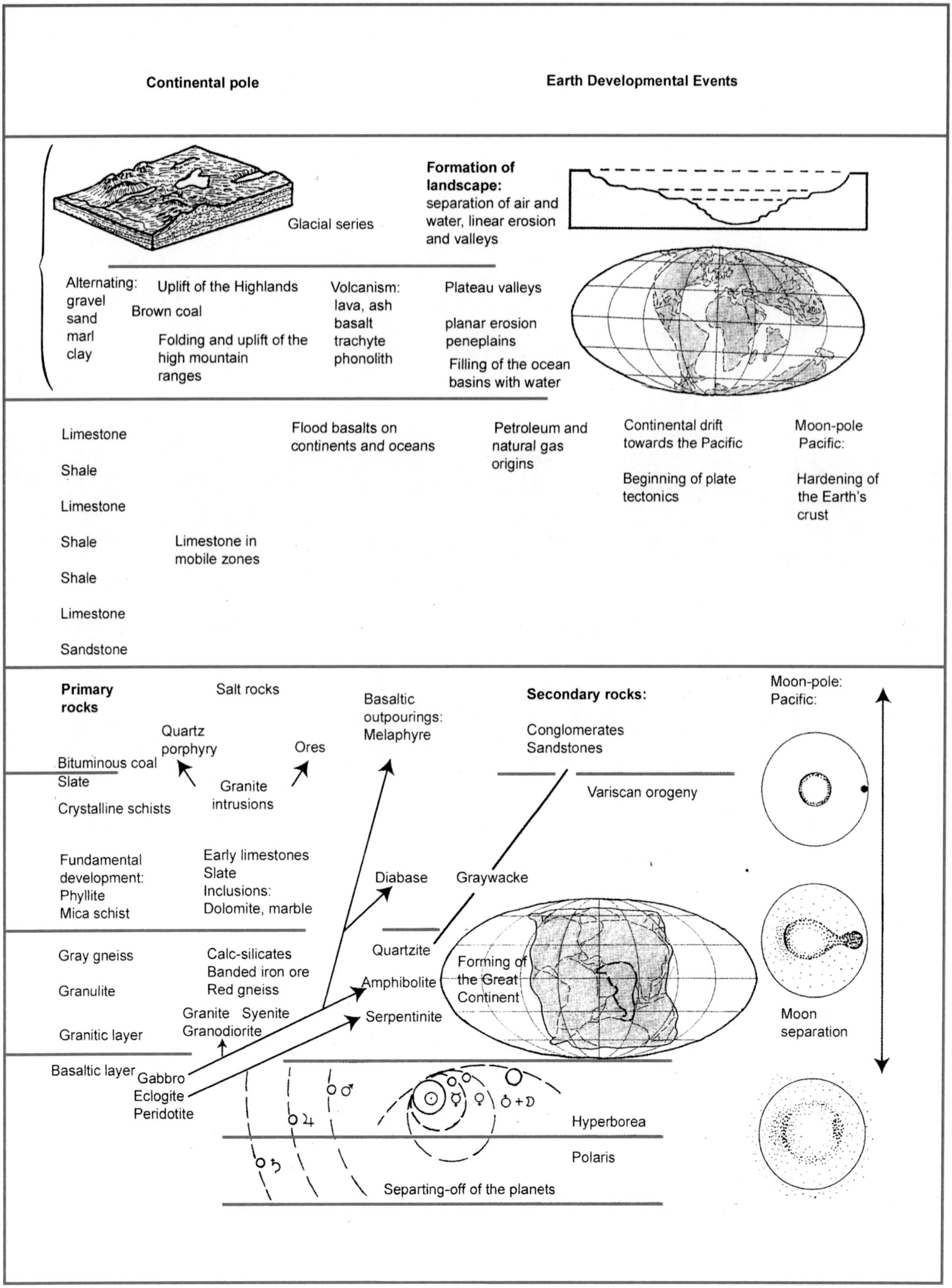

Continental pole
Earth Developmental Events
Glacial series
Formation of landscape: separation of air and water, linear erosion and valleys
Alternating: gravel sand marl clay
Uplift of the Highlands
Brown coal
Folding and uplift of the high mountain ranges
Volcanism: lava, ash basalt trachyte phonolith
Plateau valleys
planar erosion peneplains
Filling of the ocean basins with water
Limestone
Shale
Limestone
Shale
Shale
Limestone
Sandstone
Limestone in mobile zones
Flood basalts on continents and oceans
Petroleum and natural gas origins
Continental drift towards the Pacific
Beginning of plate tectonics
Moon-pole Pacific:
Hardening of the Earth's crust
Primary rocks
Salt rocks
Quartz porphyry
Ores
Bituminous coal
Slate
Crystalline schists
Granite intrusions
Basaltic outpourings: Melaphyre
Secondary rocks:
Conglomerates Sandstones
Variscan orogeny
Moon-pole: Pacific:
Fundamental development: Phyllite Mica schist
Early limestones Slate Inclusions: Dolomite, marble
Diabase
Graywacke
Gray gneiss
Granulite
Granitic layer
Calc-silicates Banded iron ore Red gneiss
Granite Syenite Granodiorite
Quartzite
Amphibolite
Serpentinite
Forming of the Great Continent
Moon separation
Basaltic layer
Gabbro
Eclogite
Peridotite
Hyperborea
Polaris
Separting-off of the planets

How can granite, the primary rock of the Earth's continents, form from the basalt that lacks the necessary ingredients?

This question drives the book's writing, which describes a step-wise approach to comprehending this conundrum in easy-to-understand terms. Part I covers the evolution of granite, which is seen in its various forms wherever the substance occurs. Part II is about the formation of the Earth, from the formation of the solar system, the appearance of basalt, and the ordering of time in Earth's history. Part III gives voice to the various epochs through Earth's formation, including the appearance of the nature kingdoms and humanity. The book is illustrated with one-hundred thirty-nine color photographs, charts, and diagrams to bring the reader closer to the answer to the question posed by the title.

DANKMAR BOSSE was born in Weimar, Germany in 1940. He studied geology, paleontology, and mineralogy at the Freiberg University of Mining and Technology in Saxony (East Germany) and worked for twenty-one years at the Central Geological Institute in East Berlin, mainly on regional geology and middle European tectonics. Since then, he has been independently active with seminars, lectures, and anthroposophical research work in all earth-scientific fields with numerous research excursions. His numerous books on geology, mineralogy, crystallography, and the Earth's history can be found in German and English editions.

Made in United States
Orlando, FL
04 November 2026

90211665R00074